$$\frac{\partial^2 f}{\partial x^2} + \frac{\partial^2 f}{\partial y^2} + \frac{\partial^2 f}{\partial z^2} = 0$$

3D FD on Laplacian

for Computational Electromagnetics

in MATLAB

$$\nabla^2 f(x, y, z) = 0$$

Mohammad Nuruzzaman

Electrical Engineering Department
King Fahd University of Petroleum & Minerals
Dhahran, Saudi Arabia

CreateSpace
4900 LaCross Road, North Charleston
SC 29406, USA
www.createspace.com

Dr. Mohammad Nuruzzaman
Electrical Engineering Department
King Fahd University of Petroleum and Minerals
KFUPM BOX 1286
Dhahran 31261, Saudi Arabia
Email: nzaman@kfupm.edu.sa, nzaman@ymail.com, mzamandr@gmail.com
Skype: nzaman1769   Twitter: @nzaman00
Web Link: http://faculty.kfupm.edu.sa/EE/NZAMAN/

ISBN-13: 978-1533524799
ISBN-10: 1533524793

Printed in the United States of America

This book is printed on acid-free paper.

To my parents
Mohammad Shamsul Haque & Nurbanu Begum

Other titles by the author:

1. M. Nuruzzaman, *"Solving Electronic Circuits in MATLAB and SIMULINK"*, October, 2015, CreateSpace, South Carolina.
2. M. Nuruzzaman, *"Control System Analysis & Design in MATLAB and SIMULINK"*, June, 2014, Lulu Press, Inc., North Carolina.
3. M. Nuruzzaman, *"Finite Difference Fundamentals in MATLAB"*, July, 2013, CreateSpace, South Carolina.
4. M. Nuruzzaman, *"Digital Image: Theories, Algorithms, and Applications"*, June, 2012, CreateSpace, Washington.
5. M. Nuruzzaman, *"Digital Audio Fundamentals in MATLAB"*, July, 2010, CreateSpace, California.
6. M. Nuruzzaman, *"Modern Approach to Solving Electromagnetics in MATLAB"*, January, 2009, BookSurge Publishing, Charleston, South Carolina.
7. M. Nuruzzaman, *"Signal and System Fundamentals in MATLAB and SIMULINK"*, July, 2008, BookSurge Publishing, Charleston, South Carolina.
8. M. Nuruzzaman, *"Electric Circuit Fundamentals in MATLAB and SIMULINK"*, October, 2007, BookSurge Publishing, Charleston, South Carolina.
9. M. Nuruzzaman, *"Technical Computation and Visualization in MATLAB for Engineers and Scientists"*, February, 2007, AuthorHouse, Bloomington, Indiana.
10. M. Nuruzzaman, *"Digital Image Fundamentals in MATLAB"*, September, 2005, AuthorHouse, Bloomington, Indiana.
11. M. Nuruzzaman, *"Modeling and Simulation in SIMULINK for Engineers and Scientists"*, January, 2005, AuthorHouse, Bloomington, Indiana.
12. M. Nuruzzaman, *"Tutorials on Mathematics to MATLAB"*, March, 2003, AuthorHouse, Bloomington, Indiana.

Preface

Despite limitations finite difference (FD) is a fantastic tool to solve many scientific and engineering problems. Had we had close form solutions to all real world problems, we would not have thought about the finite difference. The technique is abundantly exercised in two dimensional (2D) engineering problems. Higher dimensionality of electromagnetics frequently requires three dimensional finite difference (3D FD) which is not very common like 2D counterpart perhaps owing to the complexity involved. Electromagnetic systems are three dimensional to a large extent consequently 3D FD finds a wide variety of applications in the field. The text "*3D FD on Laplacian for Computational Electromangetics in MATLAB*" primarily focuses on solving Laplace equation applying three dimensional finite difference in Cartesian system. Simulation emphasis has been given in MATLAB, a popular computer simulation platform for technical problems.

Incredibly compact electronic circuitry are being evolved every year which is why consumers are getting new gadgets, touch/smart phones, not to mention study tools using computer. Since MATLAB's inception (whose elaboration is matrix laboratory) during 1990s, it has been incorporated into many science and engineering branches. Embedded functions in MATLAB are plenty in number and do not call for reprogramming, nor does require lumbering compiling often encountered in base language such as in FORTRAN or C. Research and development (R &D) of electromagnetics indeed involve simulation tool for which the text is devised. Intrinsically MATLAB data maneuvering is in rectangular matrix or array form, 3D FD samples shape to an ordinary 3D array after finding Laplace equation solution hence perfect platform to compute and visualize the solution. Although laser sharp focus is on the solution, application of the 3D FD solution is well demonstrated to electromagnetic systems. Analyzing convenience of the topic using 3D FD reveals one interesting fact, unsolvable analytical solution or compounded boundary condition is no exception which is not lenient in traditional harmonic or variable separation method.

Chapter 1 presents a brief introduction to MATLAB's getting-started features. Without the discourse of two dimensional finite difference (2D FD) extension to the 3D counterpart is not well understood for this reason smooth changeover from 2D to 3D is addressed in chapter 2. One regrettable fact is there is no 3D FD Laplace equation solving tool available in MATLAB. The author devised a function file which solves the Laplace equation using 3D FD. What is so complicated about the 3D Laplace equation solution? With a change in the boundary condition, new numeric circumstance appears and the conditions may engage impulse, linear, planar or other form therefore the function has to work out multitude of conditions which are addressed deliberately by the author written function **lap3d** in chapter 3. Chapter 4 merely demonstrates analysis, visualization, and application of Laplace equation solution tested on various electromagnetic systems. Emphasis has been given on Cartesian system and rectangular box based space while implementing the 3D FD. Pertinent tools are exercised to get the reader hand-on experience to finished solutions. Nonetheless appendices A through D explain 3D FD system related coding, function, or embedded graphing tool to the context of MATLAB.

My words of acknowledgement are due to the King Fahd University of Petroleum and Minerals (KFUPM). I am especially appreciative of library facilities, finite difference system reading materials, and MATLAB software that I received from the university.

Mohammad Nuruzzaman

Acknowledgements

I sincerely thank the following for their inputs, encouragements, comments, and suggestions about my MATLAB/SIMULINK text's development:

Mohsin M. Jamali (Department of Electrical Engineering and Computer Science, The University of Toledo, Ohio, USA), Rama VenKat (Electrical and Computer Engineering, University of Nevada, Las Vegas, Nevada, USA), Gilberto E. Urroz (Department of Civil & Environmental Engineering, Utah Water Research Laboratory, Utah State University, Utah, USA), G. P. Rangaiah (Department of Chemical and Environmental Engineering, National University of Singapore, Singapore), Flaminio Squazzoni (Department of Social Sciences, University of Brescia, Italy), G. N. Reddy (Department of Electrical Engineering, Lamar University, Texas, USA), Kashif Javaid (Audio Applications, National Semiconductor), Bahram Shahian (Electrical Engineering, California State University, Long Beach, USA), Lihong Li (Department of Engineering Science and Physics, Staten Island College, New York, USA), Rollins Turner (Department of Computer Science and Engineering, University of South Florida, USA), Sally Hawkins (Computer Science and Engineering, University of Nebraska-Lincoln, Nebraska, USA), Clinton Fookes (School of Engineering Systems, Faculty of Built Environment and Engineering, Queensland University of Technology, Brisbane, Australia), Swamy Jagannatham (University of California: Extension School, Los Angeles, USA), Ahmed Mabrouk (Computer Science Department, Islamic University of Malaysia, Malaysia), Jimmy Thomas (School of Biological Sciences, University of Canterbury, New Zealand), Irineu Antunes Junior (CECS, UFABC, Brazil), Kamel Alboaouh (Alumni, King Fahd University of Petroleum and Minerals, Saudi Arabia), John Bofarull Guix (London Metropolitan, UK), Ronald Tangelder (www.freenet.com, Germany), Nopparat Seemuang (Department of Mechanical Engineering, The University of Sheffield, UK), Andy Blanco (Southern California Institute of Technology, USA), Jonathan Ibera (Southern California Institute of Technology, Anaheim, USA), Young Sup Lee (Department of Embedded Systems Engineering, Incheon National University, South Korea), and Ismat Al-Dmour (Computer Science and Information Technology, Al-Baha University, Saudi Arabia).

Table of Contents

Appendices

Chapter 1

Introduction to MATLAB

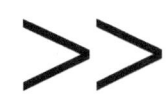

MATLAB is a computing software, which provides the quickest and easiest way to compute scientific and technical problems and visualize the solutions. As worldly standard for simulation and analysis, engineers, scientists, and researchers are becoming more and more affiliated with MATLAB. The general questionnaires about MATLAB before one gets started with are contents of this chapter. Much of MATLAB computing approach presupposes that the element to be handled is a vector or matrix. Our highlight covers the following:

- ♦ MATLAB features available at the command window
- ♦ Getting started in MATLAB starting from scratch
- ♦ Frequently encountered questions about MATLAB environment
- ♦ Relevant introductory topics and forms of assistance in MATLAB

1.1 What is MATLAB?

MATLAB is mainly a scientific and technical computing software whose elaboration is matrix laboratory. Command prompt of MATLAB (>>) provides an interactive system. In the workspace of MATLAB, most data element is dealt as a matrix without dimensioning. The package is incredibly advantageous for matrix-oriented computing. MATLAB's easy-to-use platform enables us to compute and manipulate matrices, perform numerical analyses, and visualize different variety of one/two/three dimensional graphics in a matter of second or seconds without conventional programming as conducted in FORTRAN, PASCAL, or C.

1.2 MATLAB's opening window features

If you do not have MATLAB installed in your personal computer, contact MathWorks (owner and developer, www.mathworks.com) for the installation CD. If you know how to get in MATLAB and its basics, you may skip the chapter. Assuming the package is installed in your system, run MATLAB from the Start of the Microsoft Windows. Let us get familiarized with MATLAB's opening window features. Figure 1.1(a) shows a typical firstly opened MATLAB window. Depending on desktop setting or MATLAB version, your MATLAB window may not look like figure 1.1(a) but descriptions of the features by and large are appropriate.

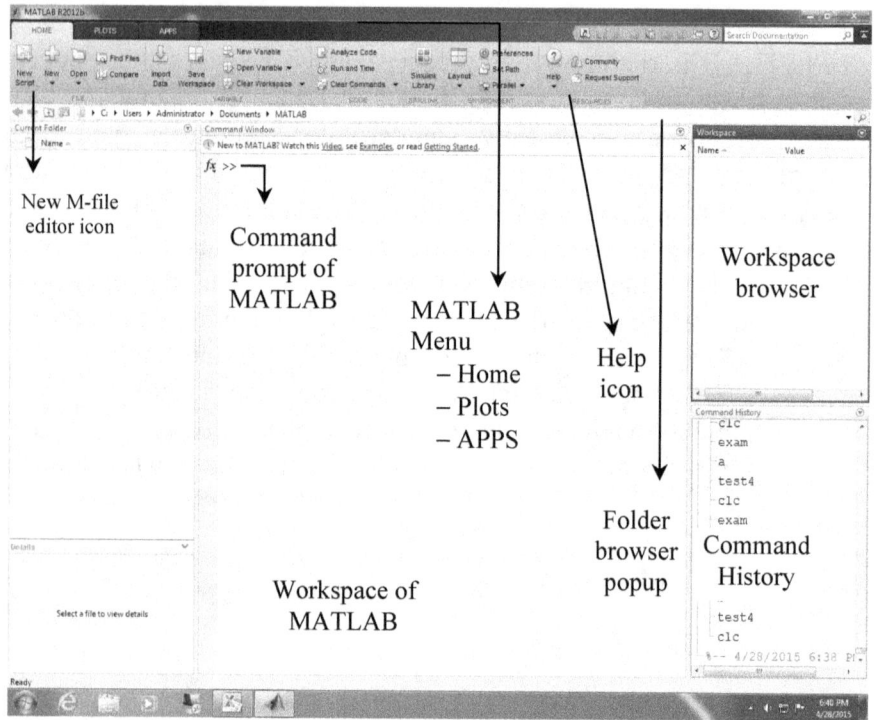

Figure 1.1(a)   Typical features of MATLAB's firstly opened window

✦ Command prompt of MATLAB

Command prompt means that you tell MATLAB to do something from here (>> in figure 1.1(a)). As an interactive system, MATLAB responds to user through this prompt. MATLAB cursor will be blinking after >> prompt once you open MATLAB i.e. MATLAB is ready to take your commands. To enter any command, type executable MATLAB statements from keyboard and to execute that, press Enter key (symbol ↵ for 'Hit the Enter Key' operation).

✦ MATLAB Menu

MATLAB is accompanied with three submenus namely HOME, PLOTS, and APPS. Each submenu has its own features. Use the mouse to click different submenus and their brief descriptions are as follows:

Submenu HOME: This is basically the firstly opened default window i.e. figure 1.1(a). It allows us to open a new script or M-file, model, or Graphical User Inter-face (GUI) layout maker, open a file which was saved before, load a saved workspace, import data from a file, save the workspace variables, set the required path to execute a file, print the workspace, and keep provision for changing the command window property.

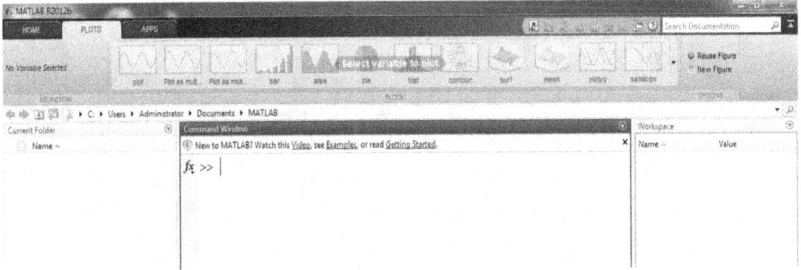

Figure 1.1(b) Submenu PLOTS

Submenu PLOTS: The second submenu PLOTS (figure 1.1(b)) includes some of available graphical tools embedded in MATLAB. MATLAB has vast graphics supports, above is just sample on that.

Submenu APPS: MATLAB has numerous built-in libraries which conduct specific discipline oriented simulations, these libraries are called Toolbox or APPS. The submenu APPS (figure 1.1(c)) just displays some of many APPS available. For example in its own menu bar you find Curve Fitting, Optimization, etc.

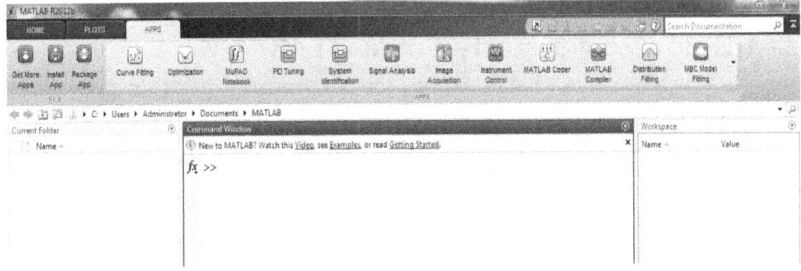

Figure 1.1(c) Submenu APPS

✦ MATLAB workspace

Workspace (figure 1.1(a)) is the platform of MATLAB where one executes MATLAB commands. During execution of commands, one may have to deal with some input and output variables. These variables can be one-dimensional array, multi-dimensional array, characters, symbolic objects, etc. Again to deal with graphics window, we have texts, graphics, or object handles. Workspace holds all those variables or handles for you. As a subwindow of figure 1.1(a), its browser exhibits the types or properties of those variables or handles. If the browser is not found in the opening window of MATLAB, click the Layout down Workspace in the HOME bar to bring the subwindow (figure 1.1(d)).

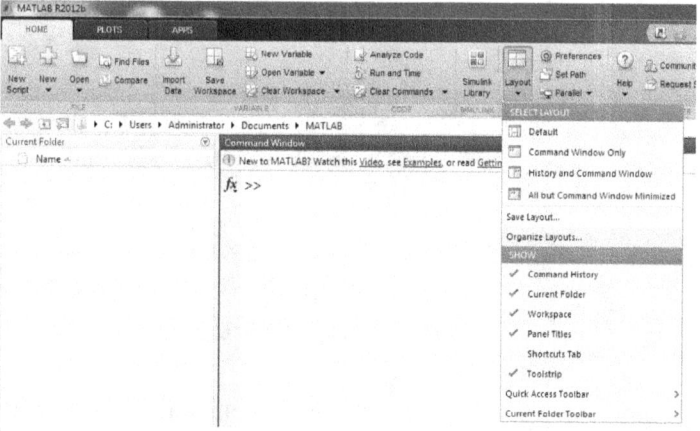

Figure 1.1(d) Layout pulldown menu under HOME

✦ MATLAB command history

There is a subwindow in figure 1.1(a) called Command History which holds all previously used commands at command prompt. Depending on desktop setting, it may or may not appear during the opening of MATLAB. If it does not, click the Command History from figure 1.1(d) under the Layout. Used commands of this window you may copy-paste in an editor for later use.

1.3 How to get started in MATLAB?

New MATLAB users face a common question how to get started in MATLAB? This tutorial is for beginners in MATLAB. Here we address the terms under the following bold headings.

✦ How can I enter a vector/matrix?

The first step is the user has to be in the command window of MATLAB. Look for the command prompt >> in the command window. Row or column matrices are termed as vectors. We intend to enter the row matrix

R=[2   3   4   −2   0] into the workspace of MATLAB. Type the following from the keyboard at the command prompt:

>>R=[2 3 4 -2 0]   ← Arial font set for executable commands i.e. R⇔R

There is one space gap between two elements of the matrix R but no space gap at the edge elements. All elements are placed under the [ ]. Press Enter key after the third brace] from the keyboard and we see

R =

2   3   4   -2   0

>>   ← command prompt is ready again

It means we assigned the row matrix to the workspace variable R. Whenever we call R, MATLAB understands the whole row matrix. Matrix R is having five elements. Even if R had 100 elements, it would understand the whole matrix that is one of many appreciative features of MATLAB. Next we wish to enter the column matrix C=$\begin{bmatrix} 7 \\ 8 \\ 10 \\ -11 \end{bmatrix}$. Again type the following from the keyboard at the blinking cursor:

>>C=[7;8;10;-11] ↵   you will see (↵ means 'Press the Enter Key'),

C =

7
8
10
−11

>>   ← command prompt is ready again

This time we also assigned the column matrix to the workspace variable C. For the column matrix, there is one semicolon ; between two consecutive elements of the matrix C but no space gap is necessary. As another option, the matrix C could have been entered by writing C=[7 8 10 -11]'. The operator ' of keyboard is matrix transposition operator in MATLAB. As if you entered a row matrix but at the end just the transposition operator ' is attached. After that the rectangular matrix A=$\begin{bmatrix} 20 & 6 & 7 \\ 5 & 12 & -3 \\ 1 & -1 & 0 \\ 19 & 3 & 2 \end{bmatrix}$ is to be entered:

>>A=[20 6 7;5 12 -3;1 -1 0;19 3 2] ↵   you will see,

A =

20   6   7
5   12   -3
1   -1   0
19   3   2

Two consecutive rows of A are separated by semicolon ; and consecutive elements in a row are separated by one space gap. Instead of typing all elements in a row, one can type the first row, press Enter key, the cursor blinks in the next line, type the second row, and so on.

✦ How can I use colon and semicolon operators?

Operators semicolon ; and colon : have special significance in MATLAB. Most MATLAB statements and M-file programming use these two operators almost in every line. Generation of vectors is easily performed by colon operator no matter how many elements we need. Let us carry out the following at command prompt to see the importance of colon operator:

```
>>A=1:4 ↵          you will see,
```

```
A =

    1   2   3   4        ← We created a vector A or row matrix
                             where A=[1   2   3   4]
```

Let us interact with MATLAB by the following commands:

```
>>R=1:3:10 ↵          you will see,
```

```
R =

    1   4   7   10       ← We created a vector or row matrix R whose elements form an
                             arithmetic progression with first element 1, last element 10,
                             and common difference or increment 3
```

Vector with decrement can also be generated:

```
>>C=[0:-2:-10]' ↵          you will see,
```

```
C =

     0
    -2
    -4          ← We created a vector or column matrix C whose
    -6             consecutive elements have the decrement 2 with the
    -8             first element 0 and the last element −10
   -10
```

MATLAB is also capable of producing vectors whose elements are decimal numbers. Let us form a row matrix R whose first element is 3, last element is 6, and increment is 0.5 which we accomplish as follows:

```
>>R=3:0.5:6 ↵     you will see,
```

```
R =
    3.0000   3.5000   4.0000   4.5000   5.0000   5.5000   6.0000
```

Then, what is the use of semicolon operator? Append a semicolon at the end in the last command and execute that:

```
>>R=3:0.5:6; ↵     you will see,
>>                                      ← Assignment is not shown
```

Type R at the command prompt and press Enter:

```
>>R ↵
```

```
R =
    3.0000   3.5000   4.0000   4.5000   5.0000   5.5000   6.0000
```

It indicates that the semicolon operator prevents MATLAB from displaying the contents of the workspace variable R.

✦ How can I call an embedded or a built-in MATLAB function?

In MATLAB, thousands of M-files or built-in function files are embedded. Knowing descriptions of the function, numbers of input and output arguments, and nature of the arguments is mandatory in order to execute a built-in function. Let us start with a simplest example. We intend to find $\sin x$ for $x = \dfrac{3\pi}{2}$ which should be -1. The MATLAB counterpart (appendix A) of $\sin x$ is sin(x) where x can be any real or complex number in radians and can be a matrix too. The angle $\dfrac{3\pi}{2}$ is written as 3*pi/2(π is coded by pi) and let us perform it as follows:

```
>>sin(3*pi/2) ↵

ans =
     -1
```

By default the return from any function is assigned to workspace ans. If you wanted to assign the return to S, you would write S=sin(3*pi/2);.

As another example, let us factorize the integer 84 (84=2×2×3×7). The MATLAB built-in function factor finds the factors of an integer and the implementation is as follows:

```
>>f=factor(84) ↵

f =
     2   2   3   7
```

Output of the factor is a row matrix which we assigned to workspace f in fact the f can be any user-given name. Thus you can call any other built-in function from the command prompt provided that you have the knowledge about the calling of inputs to and outputs from the function.

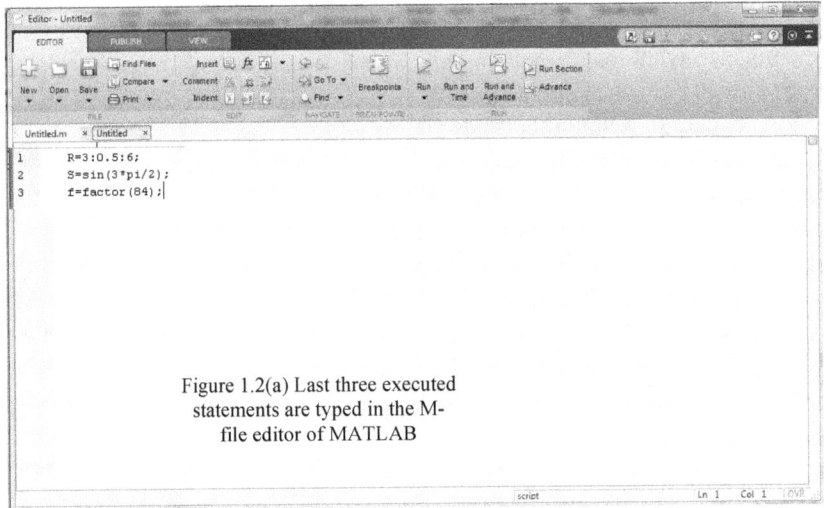

Figure 1.2(a) Last three executed statements are typed in the M-file editor of MATLAB

♦ **How can one open and execute an M-file or a script file?**

This is the most important start up for the beginners. An M-file can be regarded as a text or script file. A collection of executable MATLAB statements are the contents of an M-file. Ongoing discussion made you familiarize with entering a matrix, computing a sine value, and factorizing an integer. These three executions took place at the command prompt. They can be executed from an M-file as well. This necessitates opening the M-file editor. Referring to figure 1.1(a), you find the icon for **New Script** in the upper left corner of the window, click it to see the new untitled script editor. After opening the new script editor, we typed the last three executable statements in the untitled file as shown in figure 1.2(a). The next step is to save the untitled file by clicking the Save icon of figure 1.2(a). Figure 1.2(b) presents the File Save dialog window. We typed the file name as **test** (can be any name of your choice) in the slot of File name in the window. The script file has the extension .**m** but we do not type .**m** only the file name is enough. After saving the file, let us move on to MATLAB command prompt and conduct the following:

>>test ↵
>>   ← command prompt is ready again

It indicates that MATLAB executed the M-file by name **test** and is ready for next command. We can check calling the assignees whether the previously performed executions occurred exactly as follows:

>>R ↵

R =
      3.0000   3.5000   4.0000   4.5000   5.0000   5.5000   6.0000
>>S ↵   >>f ↵

S =   f =
      -1   2   2   3   7

This is what we found before. Thus one can run any executable statements in script file. The reader might ask in which folder or path the **test** was saved. Figure 1.1(a) shows the bar (down the menu bar in the upper portion) for Current Folder which is C:\Users\Administrator\ Documents\MATLAB. That is the location of your file. If you want to save the script file in other

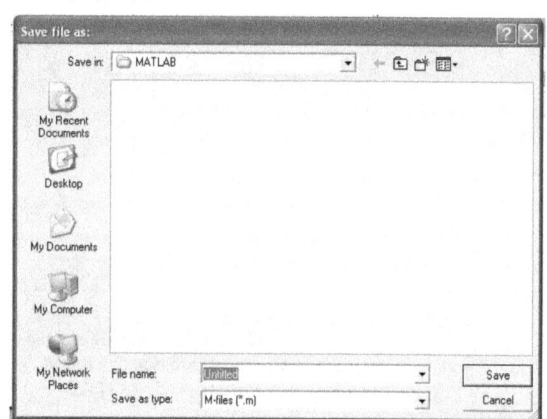

Figure 1.2(b) Save dialog window for naming the M-file

folder or directory, change path by clicking the path browser icon before saving the file. When you call the **test** or any other file from the command prompt, the prompt must be in the same directory where the file is in or its path must be defined to MATLAB.

✦ What are input and output arguments of a function file?

MATLAB is a collection of thousands of script files. Some files are executed without any return and some return results which are called function files (appendix D). You have seen the use of function sin(x) before, which has one input argument x. The statement test(x,y) means that the test is a function file which has two input arguments - x and y. Again the test(x,y,z) means the test is a function file which needs three input arguments - x, y, and z. Similar style also follows for the return but under the third brace. The [a,b]=test(x,y) means there are two output arguments from the test which are a and b and the [a,b,c]=test(x,y) means three returns from the test which are a, b, and c.

✦ How can I plot a graph?

MATLAB is very convenient for plotting different sorts of graphs. The graphs are plotted either from mathematical expression or from data. Let us plot the function $y = -2\sin 2x$. MATLAB function ezplot plots y versus x type graph taking the expression as its input argument. MATLAB code (appendix A) for the $-2\sin 2x$ is -2*sin(2*x). The functional code is input argumented by using single inverted comma hence we conduct the following at command prompt:

>>ezplot('-2*sin(2*x)') ⏎

Edit plot Icon Zoom In and Out Icons

Figure 1.2(c) Graph of $-2\sin 2x$ versus x

Figure 1.2(c) presents the outcome from above execution. The window in which the graph is plotted is called MATLAB figure window. Any graphics is plotted in the figure window, which has its own menu (such as File, Edit, etc) as shown in figure 1.2(c).

1.4 Some queries about MATLAB environment

Users need to know the answers to some questions when they start working in MATLAB. MATLAB environment related some queries are presented in the sequel.

⊟ **How to change the numeric format?**

When you perform any computation at the command prompt, the output is returned up to four decimal display due to short numeric format which is the default one. There are other numeric formats too. To reach the numeric format dialog box, the clicking operation sequence is HOME ⇒ Preferences ⇒ Command Window ⇒ Text Display ⇒ Numeric Format (select from the popup menu e.g. long).

⊟ **How to change the font or background color settings?**

One might be interested to change the background color or font color while working in the command window. The clicking sequence is HOME ⇒ Preferences ⇒ MATLAB. You find Desktop Tool Colors in the right half window, uncheck that and select any Text or Background Color from the popup.

⊟ **How to delete some/all variables from the workspace?**

In order to delete all variables present at the workspace, the clicking sequence is HOME ⇒ Clear Workspace (figure 1.1(a)). If you want to delete a particular workspace variable, select the concern variable by using the mouse pointer in the workspace browser (assuming that it is open

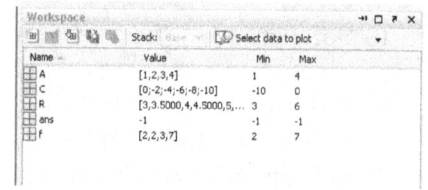

Figure 1.2(d) Workspace browser displays variable information

like the figure 1.2(d)) and then rightclick ⇒ delete.

⊟ **How to clear workspace but not the variables?**

Once you conduct some sessions at the command prompt, monitor screen keeps all interactive sessions. You can clear the screen contents without removing the variables by command clc or performing the clicking operation HOME ⇒ Clear Commands (figure 1.1(a)).

⊟ **How to know the current path?**

In the upper portion of figure 1.1(a), the Current Folder bar is located that indicates in which path the command prompt is or execute cd (abbreviation for the current directory) at the command prompt.

☐ How to see different variables at the workspace?

There are two ways of viewing - either use the command who or look at the workspace browser (like figure 1.2(d)) which exhibits information about workspace variables for example R is the name of variable which holds some values. One can view, change, or edit the contents of a variable by doubleclicking the concern variable situated at workspace browser.

☐ How to enter a long command line?

MATLAB statements can be too long to fit in one line. Giving a break in the middle of a statement is accomplished by the ellipsis (three dots are called ellipsis). We show that considering the entering of vector x=[1:3:10] as follows:

```
>>x=[1:3: ... ↵
        10] ↵
x =
      1   4   7   10
```

Typing takes place in two lines and there is one space gap before the ellipsis.

☐ Editing at the command prompt

This is advantageous specially for those who work frequently in the command window without opening a script file. Keyboard has different arrow keys marked by ← ↑ → ↓. One may type a misspelled command at the command prompt causing error message to appear. Instead of retyping the entire line, press uparrow (for previous line) or downarrow (for next line) to edit the MATLAB statement. Or you can reexecute any past statement this way. For example we generated a row vector 1 through 10 with increment 2 and assigned the vector to x. The necessary command is x=1:2:10. Mistakenly you typed x+1:2:10. The response is as follows:

```
>>x+1:2:10 ↵
??? Undefined function or variable 'x'.
```

You discovered the mistake and want to correct that. Press ↑ key to see,

```
>>x+1:2:10
```

Edit the command going to the + sign by using the left arrow key or mouse pointer. At the prompt, if you type x and press ↑ again and again, you see used commands that start with x.

☐ Saving and loading data

User can save workspace variables or data in a binary file having the extension .mat. Suppose you have the matrix $A = \begin{bmatrix} 3 & 4 & 8 \\ 0 & 2 & 1 \end{bmatrix}$ and wish to save

A in a file by the name data.mat. Let us carry out the following:

```
>>A=[3 4 8;0 2 1]; ↵          ← Assigning the A to A
```

Now move on to the workspace browser (figure 1.2(d)) and you see the variable A including its information located in the subwindow. Bring mouse pointer on A, rightclick the mouse, and click the Save As. The Save dialog window appears and type only data (not data.mat) in the slot of File name. If it is necessary, you can save all workspace variables by using the same action but clicking HOME ⇒ Save Workspace (figure 1.1(a)). One retrieves the data file by clicking HOME ⇒ Import Data (figure 1.1(a)). Another option is use the command load data at the command prompt.

⊡ **How to delete a file from the command prompt?**

Let us delete just mentioned data.mat by executing the command delete data.mat at the command prompt.

⊡ **How to see the data held in a variable?**

Figure 1.2(d) presents some variable information in which you find R. Doubleclick the R or your variable in the workspace browser and find the matrix contents of R in a data sheet.

1.5 How to get help?

Help facilities in MATLAB are plentiful. One can access to information about a MATLAB function in a variety of ways. Command help finds the help of a particular function file. You are familiar with the function sin(x) from earlier discussion and can have command prompt help regarding the sin(x) as follows:

```
>>help sin ⌐                          ← Function name without the argument
sin   Sine of argument in radians.
   sin(X) is the sine of the elements of X.

   See also asin, sind.
      ⋮
```

One disadvantage of this method is the user has to know the exact file name of a function. For a novice this facility may not be appreciative.

Casually we know partial name of a function or try to check whether any function exists by that name. Suppose we intend to see whether any function by the name finitediff exists. Machine executes that through the intermediacy of command lookfor (no space gap between look and for, the reader needs to be patient because searching from the whole MATLAB package may take at least two minutes or longer depending on your computer speed):

```
>>lookfor finitediff ⌐
```

```
finitedifferences          - computes finite-difference derivatives.
parfinitedifferences        - computes finite-difference derivatives in parallel.
```

The last return is showing all possible functions bearing the file name finitediff or having the file name finitediff partly. Now the command help can be conducted to go through a particular one for example the first one is finitedifferences and execute help finitedifferences to see its description at the command prompt:

```
>>help finitedifferences ⏎

FINITEDIFFERENCES computes finite-difference derivatives.

This helper function computes finite-difference derivatives of the objective
and constraint functions.

[gradf,cJac,NEWLAMBDA,OLDLAMBDA,s] = FINITEDIFFERENCES(xCurrent, ...
    xOriginalShape,funfcn,confcn,lb,ub,fCurrent,cCurrent, ...
    XDATA,YDATA,DiffMinChange,DiffMaxChange,typicalx,finDiffType, ...
    variables,LAMBDA,NEWLAMBDA,OLDLAMBDA,POINT,FLAG,s, ...
    varargin)
computes the finite-difference gradients of the objective and
constraint functions.
```

⋮

More descriptive texts are displayed, you may view those in the command window of MATLAB.

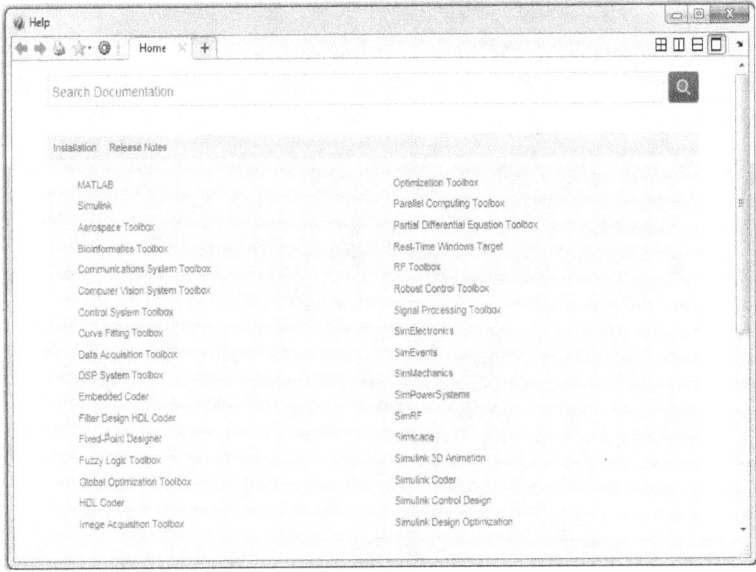

Figure 1.3(a) General Help window of MATLAB

In order to have window form help, click the Help icon (i.e. ?) of figure 1.1(a) and MATLAB responds with the Help window of figure 1.3(a). As the figure shows, help is available based on the contents. If you have some search word on MATLAB, search that through the **Search Documentation** of figure 1.3(a). This help form is better when one navigates MATLAB's capability not looking for a particular function.

Hidden algorithm or mathematical expression is often necessary whose assistance we can have through the search option from MathWorks Website provided that your PC is connected to internet.

However we close the introductory discussion on MATLAB with this.

Chapter 2

Finite Difference in Two and Three Dimensions

Despite three dimensional finite difference (3D FD) is the extension of two dimensional counterpart (2D FD), extra concepts or terminologies are essential to implement 3D FD problems owing to added complexity. In one and two dimensional FD, basis vectors are organized as row/column and rectangular matrices respectively whereas in three dimensional counterpart recurrently three dimensional array grid points are involved which make the computing to some extent complicated. A new-fangled understanding needs to be developed for which we highlight the following:

- ♦ ♦ Mathematical FD model of two and three variable expressions
- ♦ ♦ FD grid point links of two and three variable functions
- ♦ ♦ Two and three dimensional FD data analysis in MATLAB
- ♦ ♦ Embedded tools of some mathematical operations on FD data

2.1 Model of two variable expression

Conversion of a one dimensional continuous function to its discrete counterpart is basically a row/column matrix subject to some resolution which can be extended in two dimensions. All-too-familiar two variable function $f(x,y)$ for example polynomial $x^2 - xy + 3y^2$, sinusoidal $2\sin 3(x+y)$, exponential e^{x+y}, etc are also continuous. The continuity is with respect to x, with respect to y, and with respect to $f(x,y)$.

◆ Independent variable discretization i.e. on x and y

Discretization takes place both in x and y directions which are the independent or basis variables. Let us say the continuous coordinate corresponding to any $f(x,y)$ is (x,y) where the x or y is not integer and each can assume any real number. For example x can be 1.9999, 2, or 2.0001. The x is written as $x = m\Delta x$ where the Δx is the sample period along x or horizontal direction. Similar writing is also applicable for the y i.e. $y = n\Delta y$ where the Δy is the sample period along y or vertical direction. The x or Δx has the same unit (e.g. meter) but the m does not have unit. Similar explanation is true for y, Δy, and n. The Δx and Δy are also called step size in x and y directions respectively.

Figure 2.1(a) depicts the whole strategy on the discretization. A rectangular area of size $x_0 \times y_0$ has the continuous domain variation over $0 \le x \le x_0$ and $0 \le y \le y_0$. Any point (x,y) in the

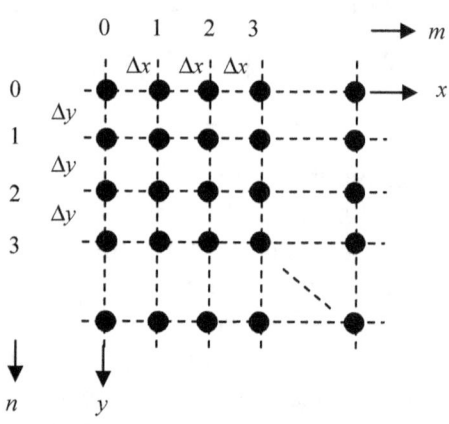

Figure 2.1(a) Link between the continuous and discrete coordinates of $f(x,y)$ due to FD

continuous domain turns to integer coordinate (m,n) in the FD domain. Every bold dot in figure 2.1(a) represents one sample point.

Suppose the continuous x and y have the intervals $0 \le x \le 8\mu m$ and $0 \le y \le 12\mu m$ respectively and $\Delta x = \Delta y = 2\,\mu m$ is chosen. The integer m varies from 0 to 4, so does n from 0 to 6 in other words the FD domain is over $0 \le m \le 4$ and $0 \le n \le 6$.

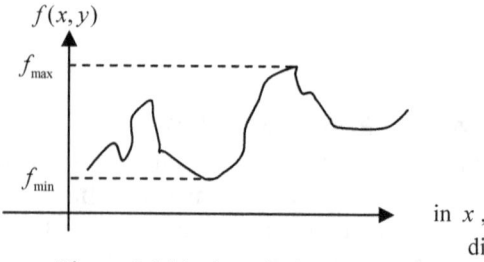

Figure 2.1(b) A typical $f(x,y)$ variation over some domain

✦ Functional discretization i.e. on $f(x,y)$

The Δx or Δy is solely related to continuous (x,y). What is about functional value resolution? Say the discretized $f(x,y)$ is labeled as $f[m,n]$. How do we derive the discrete $f[m,n]$ from continuous $f(x,y)$?

Finite difference study always works over finite range of $f(x,y)$. Any $f(x,y)$ may vary from some minimum to maximum say f_{min} to f_{max} respectively. Over a given domain you may check the functional value of $f(x,y)$ along x, y, or both directions but for sure the value is going to be between f_{min} and f_{max} as in figure 2.1(b).

Numerical example is the best for understanding a concept. Suppose the continuous $f(x,y)$ is changing from f_{min} =0.1 $\mu W / \mu m^2$ (i.e. microwatts/ micrometer^2) to f_{max} =0.9 $\mu W / \mu m^2$ over some x - y domain. Mathematically you may write the figure 2.1(b) cited variation as $0.1 \mu W / \mu m^2 \leq f(x,y) \leq 0.9 \mu W / \mu m^2$. The $f(x,y)$ at (x,y) can be any value e.g. $0.1999999 \mu W / \mu m^2$, $0.2 \mu W / \mu m^2$, or $0.20000001 \mu W / \mu m^2$ - infinite possibilities. FD does not handle infinity like situation that brings the resolution on $f(x,y)$ drawn in.

Usually we consider some predecided levels for discretization of $f(x,y)$ and divide $f_{max} - f_{min}$ by the level number minus 1 and the result is the resolution on $f(x,y)$ say Δf. For instance if we decide 5 levels, then the $f(x,y)$ with f_{min} =0.1 $\mu W / \mu m^2$ and f_{max} =0.9 $\mu W / \mu m^2$ has functional resolution Δf =(0.9-0.1)/4=0.2 $\mu W / \mu m^2$. With this probable variations of $f(x,y)$ are restricted to 0.1, 0.3, 0.5, 0.7, and 0.9 (all in $\mu W / \mu m^2$).

Inbetween $f(x,y)$ values are approximated to the nearest discretized functional value. For example $f(x,y)$=0.21 $\mu W / \mu m^2$ is considered as $0.3 \mu W / \mu m^2$. Slight functional value is lost due to this approximation but that is the price of using FD technique or discretization.

However the reconstructed functional value is computed from $\hat{f}(x,y)$ $= f_{min} + \Delta f \; f[m,n]$ with earlier $x = m\Delta x$ and $y = n\Delta y$. Note that the $\hat{f}(x,y)$, f_{min}, f_{max}, and Δf have the same unit (i.e. $\mu W / \mu m^2$) but the $f[m,n]$ does not have any unit and it is simply some integer number. In tabular form we show the link among $\hat{f}(x,y)$, Δf, and $f[m,n]$ for the numerical example as follows:

$\hat{f}(x,y)$ in $\mu W / \mu m^2$	Δf in $\mu W / \mu m^2$	$f[m,n]$ unitless
0.1		0
0.3		1
0.5	0.2	2
0.7		3
0.9		4

What can we say about the last tabular representation? Continuous functional value of $f(x,y)$ i.e. $0.1\text{-}0.9\ \mu W/\mu m^2$ is linearly transformed to integer 0-4 with resolution $\Delta f = 0.2\ \mu W/\mu m^2$ and $f_{\min} = 0.1\ \mu W/\mu m^2$.

In general functional resolution Δf is linked to $f(x,y)$ by $\Delta f = \dfrac{f_{\max} - f_{\min}}{L-1}$ where L is the user-chosen level number. The greater is the level number L, the better is the functional resolution.

Different notations are seen in the literature. For instance the $f[m,n]$ variation from 0 to 31 is mathematically written as $0 \le f[m,n] \le 31$ for 32 levels. As another notation $[0,31]$ elucidates the same.

We summarize the following for a two dimensional continuous function which has to be discretized:

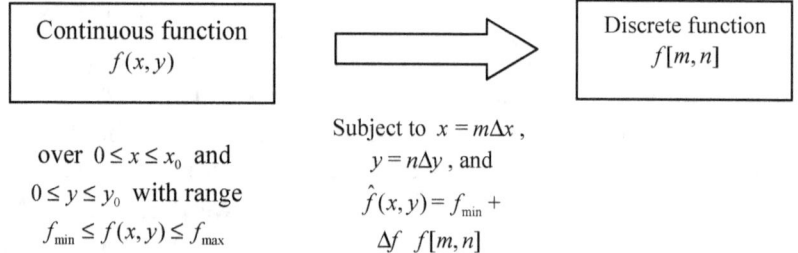

Continuous function $f(x,y)$	$\Rightarrow$	Discrete function $f[m,n]$

over $0 \le x \le x_0$ and
$0 \le y \le y_0$ with range
$f_{\min} \le f(x,y) \le f_{\max}$

Subject to $x = m\Delta x$,
$y = n\Delta y$, and
$\hat{f}(x,y) = f_{\min} +$
$\Delta f\ f[m,n]$

2.2 Rectangular matrix for 2D finite difference

Here in this section our objective is to establish a matrix link between a continuous function $f(x,y)$ and two dimensional finite difference derived $f[m,n]$.

Let us choose some $f(x,y)$ for this purpose say $f(x,y) = (x-y)^2$. The x-y domain and its sample periods are also essential so let it be $-0.5 \le x \le 1$, $-1 \le y \le 1$, $\Delta x = 0.5$, and $\Delta y = 0.5$.

For the given sample periods the x and y sample points both as a row matrix are $[-0.5\ \ 0\ \ 0.5\ \ 1]$ and $[-1\ \ -0.5\ \ 0\ \ 0.5\ \ 1]$ respectively. The sample values of $f(x,y)$ are organized as follows:

$$
\begin{array}{c}
x \rightarrow \\
y \downarrow
\end{array}
\begin{bmatrix}
f(-0.5,-1) & f(0,-1) & f(0.5,-1) & f(1,-1) \\
f(-0.5,-0.5) & f(0,-0.5) & f(0.5,-0.5) & f(1,-0.5) \\
f(-0.5,0) & f(0,0) & f(0.5,0) & f(1,0) \\
f(-0.5,0.5) & f(0,0.5) & f(0.5,0.5) & f(1,0.5) \\
f(-0.5,1) & f(0,1) & f(0.5,1) & f(1,1)
\end{bmatrix}.
$$

The samples of $f(x,y)$ evidently occupy a matrix in which the x and y coordinates are horizontally rightward and vertically downward respectively.

We followed this convention. You could choose another convention. In the first column of the matrix all we have is −0.5 in the position of x which is the first element of x sample points. Again in the first row we have only −1 in the position of y which is the first element of y sample points. Collecting only the x and only the y coordinates we get the following matrices:

$$X = \begin{bmatrix} -0.5 & 0 & 0.5 & 1 \\ -0.5 & 0 & 0.5 & 1 \\ -0.5 & 0 & 0.5 & 1 \\ -0.5 & 0 & 0.5 & 1 \\ -0.5 & 0 & 0.5 & 1 \end{bmatrix} \ and \ \ Y = \begin{bmatrix} -1 & -1 & -1 & -1 \\ -0.5 & -0.5 & -0.5 & -0.5 \\ 0 & 0 & 0 & 0 \\ 0.5 & 0.5 & 0.5 & 0.5 \\ 1 & 1 & 1 & 1 \end{bmatrix}.$$

What can we say about the last X and Y matrices? They are identical in size which is the size of $f(x,y)$ sample matrix. The X is formed from the repetition of only x point samples as a row matrix where the repetition is according to the number of y point samples. Again the Y is formed from the repetition of only y point samples as a column matrix where the repetition is according to the number of x point samples. Let us name the two matrices as grid or sample point matrices.

Having found the grid point matrices, computing of $f(x,y)$ takes place by replacing the x by X and the y by Y i.e.

$$f(X,Y) = (X-Y)^2 = \begin{bmatrix} (-0.5+1)^2 & (0+1)^2 & (0.5+1)^2 & (1+1)^2 \\ (-0.5+0.5)^2 & (0+0.5)^2 & (0.5+0.5)^2 & (1+0.5)^2 \\ (-0.5-0)^2 & (0-0)^2 & (0.5-0)^2 & (1-0)^2 \\ (-0.5-0.5)^2 & (0-0.5)^2 & (0.5-0.5)^2 & (1-0.5)^2 \\ (-0.5-1)^2 & (0-1)^2 & (0.5-1)^2 & (1-1)^2 \end{bmatrix} =$$

$$\begin{bmatrix} 0.25 & 1 & 2.25 & 4 \\ 0 & 0.25 & 1 & 2.25 \\ 0.25 & 0 & 0.25 & 1 \\ 1 & 0.25 & 0 & 0.25 \\ 2.25 & 1 & 0.25 & 0 \end{bmatrix}$$ which are the samples of $f(x,y)$.

Let us summarize the procedure in order to get sampled $f(x,y)$ from its expression in FD domain:

(a) generate x samples as a row matrix from given x variation and Δx,

(b) generate y samples as a column matrix from given y variation and Δy,

(c) repeat x samples according to the number of y samples to get X,

(d) repeat y samples according to the number of x samples to get Y, and

(e) perform $f(X,Y)$ operation to get the samples of $f(x,y)$.

❖ How to get $f[m,n]$ from sampled $f(x,y)$?

Suppose we have a rectangular matrix, it represents just samples of $f(x,y)$. In the matrix there is no information about Δx or Δy. From the sample matrix we determine the following:

f_{min} =minimum in the whole $f(x,y)$ sample matrix and

f_{max} =maximum in the whole $f(x,y)$ sample matrix.

For the ongoing example we have $f_{min}=0$ and $f_{max}=4$. The Δf needs L say $L=3$ so $\Delta f=(4-0)/(3-1)=2$. In order to get $f[m,n]$ we have to round the

$\dfrac{f(X,Y)-f_{min}}{\Delta f}$ towards its nearest integer so $f(X,Y)=\begin{bmatrix} 0.25 & 1 & 2.25 & 4 \\ 0 & 0.25 & 1 & 2.25 \\ 0.25 & 0 & 0.25 & 1 \\ 1 & 0.25 & 0 & 0.25 \\ 2.25 & 1 & 0.25 & 0 \end{bmatrix}$

provides $f[m,n]=\text{round}\{\dfrac{f(X,Y)-0}{2}\}=\text{round}\{\begin{bmatrix} 0.125 & 0.5 & 1.125 & 2 \\ 0 & 0.125 & 0.5 & 1.125 \\ 0.125 & 0 & 0.125 & 0.5 \\ 0..5 & 0.125 & 0 & 0.125 \\ 1.125 & 0.5 & 0.125 & 0 \end{bmatrix}\}=$

$\begin{bmatrix} 0 & 1 & 1 & 2 \\ 0 & 0 & 1 & 1 \\ 0 & 0 & 0 & 1 \\ 1 & 0 & 0 & 0 \\ 1 & 1 & 0 & 0 \end{bmatrix}.$

❖ How to get reconstructed $\hat{f}(x,y)$ from sampled $f[m,n]$?

The reconstructed $\hat{f}(x,y)$ is $f_{min}+\Delta f\, f[m,n]$ which becomes $\hat{f}(x,y)=$

$0+2\begin{bmatrix} 0 & 1 & 1 & 2 \\ 0 & 0 & 1 & 1 \\ 0 & 0 & 0 & 1 \\ 1 & 0 & 0 & 0 \\ 1 & 1 & 0 & 0 \end{bmatrix}=\begin{bmatrix} 0 & 2 & 2 & 4 \\ 0 & 0 & 2 & 2 \\ 0 & 0 & 0 & 2 \\ 2 & 0 & 0 & 0 \\ 2 & 2 & 0 & 0 \end{bmatrix}$ for the numerical example.

❖ How to calculate the mean square error?

In sample space the mean square error (mse) is defined as $\dfrac{1}{M\times N}\sum_x \sum_y [f(x,y)-\hat{f}(x,y)]^2$ where M and N are the numbers of columns and rows of $f(x,y)$ samples, which is tantamount to averaging all elements on matrix elements of $[f(x,y)-\hat{f}(x,y)]^2$.

For the numerical example we get the $f(x,y)-\hat{f}(x,y)$ samples as

$\begin{bmatrix} 0.25 & -1 & 0.25 & 0 \\ 0 & 0.25 & -1 & 0.25 \\ 0.25 & 0 & 0.25 & -1 \\ -1 & 0.25 & 0 & 0.25 \\ 0.25 & -1 & 0.25 & 0 \end{bmatrix}$ hence $\text{mse}=[0.25^2+(-1)^2+0.25^2+0^2+0^2+\ldots]/20=$

0.2813 where $M=4$ and $N=5$.

♦ How to get alike convention on coordinates?

Concerning the figure 2.1(a) the coordinate convention is not identical with commonly exercised one. The reason for doing so is to comply with the machine convention. Figure 2.2(a) depicts the conventional one. In the samples of $f(x,y)$ the rows are alright but the columns have to be flipped vertically in order to get the sample values complying the conventional coordinate. For the ongoing numerical example we should be having $f(x,y)$ samples or $f(X,Y)$ as

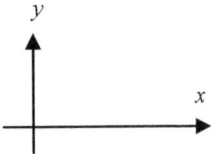

Figure 2.2(a) Common coordinate convention

$$\begin{bmatrix} 2.25 & 1 & 0.25 & 0 \\ 1 & 0.25 & 0 & 0.25 \\ 0.25 & 0 & 0.25 & 1 \\ 0 & 0.25 & 1 & 2.25 \\ 0.25 & 1 & 2.25 & 4 \end{bmatrix}$$ according to the conventional coordinate system.

2.3 Implementing two dimensional FD

Implementation on two dimensional finite difference (2D FD) is basically manipulation of rectangular matrices which is row/column matrix maneuvering in one dimensional counterpart. One dimensional functions and tools can not be applied here. Extra functional elements are required to implement 2D FD which is addressed in the sequel.

Example 1 - 2D FD from mathematical expression:

In the last section we presented one numerical example which we intend to implement. The five step procedure to compute the $f(x,y)$ samples is also applicable here.

The function **meshgrid** returns the grid point matrices (i.e. X and Y) with the syntax [user-supplied variable for X, user-supplied variable for Y]=**meshgrid**(x sample points as a row matrix, y sample points as a row matrix). After obtaining the X and Y matrices, the $f(X,Y)$ is calculated by scalar code of appendix A. Straightforwardly we implement the last section quoted numerical example as follows:

```
>>x=-0.5:0.5:1; ↵        ← x samples are assigned to x, x is user-chosen
>>y=-1:0.5:1; ↵          ← y samples are assigned to y, y is user-chosen
>>[X,Y]=meshgrid(x,y) ↵  ← Generation of grid point matrices, X⇔X , Y⇔Y ,
                                           X and Y are user-chosen

X =
        -0.5000    0    0.5000    1.0000
        -0.5000    0    0.5000    1.0000
        -0.5000    0    0.5000    1.0000
        -0.5000    0    0.5000    1.0000
        -0.5000    0    0.5000    1.0000
Y =
```

$$\begin{matrix} -1.0000 & -1.0000 & -1.0000 & -1.0000 \\ -0.5000 & -0.5000 & -0.5000 & -0.5000 \\ 0 & 0 & 0 & 0 \\ 0.5000 & 0.5000 & 0.5000 & 0.5000 \\ 1.0000 & 1.0000 & 1.0000 & 1.0000 \end{matrix}$$

Scalar code computes the samples of $f(x,y)$ which we assign to f where f is user-chosen:

>>f=(X-Y).^2 ↵

f =

$$\begin{matrix} 0.2500 & 1.0000 & 2.2500 & 4.0000 \\ 0 & 0.2500 & 1.0000 & 2.2500 \\ 0.2500 & 0 & 0.2500 & 1.0000 \\ 1.0000 & 0.2500 & 0 & 0.2500 \\ 2.2500 & 1.0000 & 0.2500 & 0 \end{matrix}$$

Two dimensional discrete function $f[m,n]$ needs $f(x,y)|_{min}$ and $f(x,y)|_{max}$ in sample space. Compute them (appendix B.5 for min/max) by the following:

>>f1=min(min(f)); ↵   ← f1⇔ $f(x,y)|_{min}$, f1 is user-chosen
>>f2=max(max(f)); ↵   ← f2⇔ $f(x,y)|_{max}$, f2 is user-chosen

Moreover we need the level number L and functional resolution Δf so execute the following:

>>L=3; df=(f2-f1)/(L-1); ↵   ← L⇔ L, df⇔ Δf, L and df are user-chosen

Hence computing for $f[m,n]$ requires:

>>fmn=round((f-f1)/df) ↵   ← fmn⇔ $f[m,n]$, fmn is user-chosen

fmn =

$$\begin{matrix} 0 & 1 & 1 & 2 \\ 0 & 0 & 1 & 1 \\ 0 & 0 & 0 & 1 \\ 1 & 0 & 0 & 0 \\ 1 & 1 & 0 & 0 \end{matrix}$$

Then the reconstructed $\hat{f}(x,y)$ samples we obtain by $f_{min}+\Delta f\ f[m,n]$:

>>f_hat=f1+df*fmn ↵   ← f_hat⇔ $\hat{f}(x,y)$, f_hat is user-chosen

f_hat =

$$\begin{matrix} 0 & 2 & 2 & 4 \\ 0 & 0 & 2 & 2 \\ 0 & 0 & 0 & 2 \\ 2 & 0 & 0 & 0 \\ 2 & 2 & 0 & 0 \end{matrix}$$

The error samples due to FD are $f(x,y)-\hat{f}(x,y)$ and obtain by:

>>e=f-f_hat ↵   ← e⇔ $f(x,y)-\hat{f}(x,y)$, e is user-chosen

e =

$$\begin{matrix} 0.2500 & -1.0000 & 0.2500 & 0 \\ 0 & 0.2500 & -1.0000 & 0.2500 \\ 0.2500 & 0 & 0.2500 & -1.0000 \end{matrix}$$

$$\begin{matrix} -1.0000 & 0.2500 & 0 & 0.2500 \\ 0.2500 & -1.0000 & 0.2500 & 0 \end{matrix}$$

Finally mean square error is computed by:

```
>>mse(e) ↵
```

```
ans =
        0.2813
```

In above the **mse** is an embedded function and calculates mean square error, input argument of which is the error rectangular matrix.

Example 2 - 2D FD on random or virtual data:

    This example addresses how to get FD data of $f(x,y)$ virtually by random variable generator of MATLAB.

    Suppose we intend to generate random samples of $f(x,y)$ each of uniform distribution within $-2V$ and $2V$ based on step sizes $\Delta x = 0.5mm$ and $\Delta y = 1mm$ over $0 \le x \le 2mm$ and $-2 \le y \le 3mm$.

    The two step sizes provide the sample numbers by $\dfrac{x_2 - x_1}{\Delta x} + 1$ and $\dfrac{y_2 - y_1}{\Delta y} + 1$ which we get as 5 and 6 in x and y directions respectively (since $x_2 = 2$ and $x_1 = 0$ from $0 \le x \le 2mm$ and $y_2 = 3$ and $y_1 = -2$ from $-2 \le y \le 3mm$). In accordance with figure 2.1(a) convention the row and column numbers in $f(x,y)$ samples are 6 and 5 respectively.

    Embedded **rand** generates uniformly distributed random numbers between 0 and 1 in the form of matrices with syntax **rand**(row number, column number). Assuming that the reader is familiar with mapping (from [0,1] to [-2,2]) which is $4 fn - 2$ and fn is the **rand** so get the machine random response for $f(x,y)$ samples in the following:

```
>>f=4*rand(6,5)-2 ↵        ← f holds f(x,y) samples, f is user-chosen
```

```
f =
        1.8005  -0.1741   1.6873  -0.3589  -1.4444
       -1.0754  -1.9260   0.9528   1.5746  -1.1889
        0.4274   1.2856  -1.2949  -1.7684  -1.2051
       -0.0561  -0.2212  -0.3772  -0.5885   0.4152
        1.5652   0.4617   1.7419   1.2527  -0.9112
        1.0484   1.1677   1.6676  -1.9606  -1.2047
```

The Δx and Δy or x - y intervals are not known from above samples.

Example 3 - 2D FD on user-supplied samples:

    When sample data of $f(x,y)$ is stored in some soft file, how do we work with that? That answer you get by going through this example.

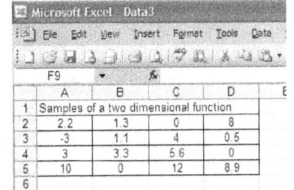

Figure 2.2(b) Samples of some $f(x,y)$ in an excel file

Figure 2.2(b) shows an excel file containing some sample data of $f(x,y)$. The file name is **Data3.xls**. Obtain the file through email link of page ii, place it in your working path of MATLAB, and execute the following:

>>f=xlsread('Data3.xls') ↵   ← f holds $f(x,y)$ samples, f is user-chosen

f =

2.2000	1.3000	0	8.0000
-3.0000	1.1000	4.0000	0.5000
3.0000	3.3000	5.6000	0
10.0000	0	12.0000	8.9000

The embedded function **xlsread** reads an excel file with the syntax xlsread(file name under quote), application of which is seen above. Note that above **f** just contains the samples of $f(x,y)$, there is no information about the step sizes or intervals.

Microsoft Excel - Data4
File Edit View Insert Format Tools Data Window Help
G13

	A	B	C	D	E	F
1	Samples of a two dimensional function				Horizontal	Vertical
2	2.2	1.3	0	8	3	0
3	-3	1.1	4	0.5	5	1
4	3	3.3	5.6	0	6	2
5	10	0	12	8.9	7	3
6						

Figure 2.2(c) Samples of figure 2.2(b) with step size information

Referring to figure 2.2(c) the step size information is provided by appending two more column data. The **Horizontal** and **Vertical** in the figure correspond to x and y sample points respectively. After obtaining the figure 2.2(c) shown file (**Data4.xls**) and placing it in working path, carry out the following:

>>V=xlsread('Data4.xls') ↵   ← V holds all data, V is user-chosen

V =

2.2000	1.3000	0	8.0000	3.0000	0
-3.0000	1.1000	4.0000	0.5000	5.0000	1.0000
3.0000	3.3000	5.6000	0	6.0000	2.0000
10.0000	0	12.0000	8.9000	7.0000	3.0000

In above V, the first four, fifth, and sixth columns are $f(x,y)$, x point, and y point samples respectively. Let us pick the samples by the following:

For $f(x,y)$ samples:
>>f=V(:,1:4) ↵   ← f holds $f(x,y)$ samples, f is user-chosen

f =

2.2000	1.3000	0	8.0000
-3.0000	1.1000	4.0000	0.5000
3.0000	3.3000	5.6000	0
10.0000	0	12.0000	8.9000

For x point samples: For y point samples:
>>x=V(:,5) ↲   >>y=V(:,6)

x =   y=
        3   0
        5   1
        6   2
        7   3

Now you may generate the grid point matrices (i.e. the X and Y matrices) by using [X,Y]=meshgrid(x,y) if it is necessary for FD related analysis like example 1.

Note that the coordinate convention of figure 2.1(a) is observed and the meshgrid works even for non uniform x or y sample points.

Example 4 - Samples in conventional coordinate system:

In this text many machine exercised problems follow figure 2.1(a) convention on 2D FD data. If the reader persists in conventional coordinate system, the function flipud (appendix B.6) can be exercised. If we consider the example 1 FD data, the conventional coordinate $f(x,y)$ data is obtained as follows:

>>f=flipud((X-Y).^2) ↲

f =
        2.2500   1.0000   0.2500   0
        1.0000   0.2500   0   0.2500
        0.2500   0   0.2500   1.0000
             0   0.2500   1.0000   2.2500
        0.2500   1.0000   2.2500   4.0000

2.4 Model of three variable expression

Three variable function $f(x,y,z)$ example can be polynomial $x^2 - xyz + 3y^2 + 5z$, sinusoidal $2\sin 3(x+2y+z)$, exponential e^{x+y-2z}, etc which are continuous too. With the added dimension continuity is with respect to x, with respect to y, with respect to z, and with respect to $f(x,y,z)$. What does it indicate? We have to deal with four quantities.

How is the discretization different from the two variable counterpart in section 2.1? Figure 2.3(a) demonstrates the discretization. The coordinate system for x and y is manifested in figure 2.3(a) too but this time we have x and y data for every single z. We arrange x and y data in a page and one page for one z data and have to visualize the figure 2.3(a) in a three dimensional context. The first page is for the $z=0$. The z also has resolution Δz. The second page refers to $z+\Delta z$, the third page refers to $z+2\Delta z$, and so on. The integer k in the z direction has the link $z=k\Delta z$ resembling to m and n of section 2.1.

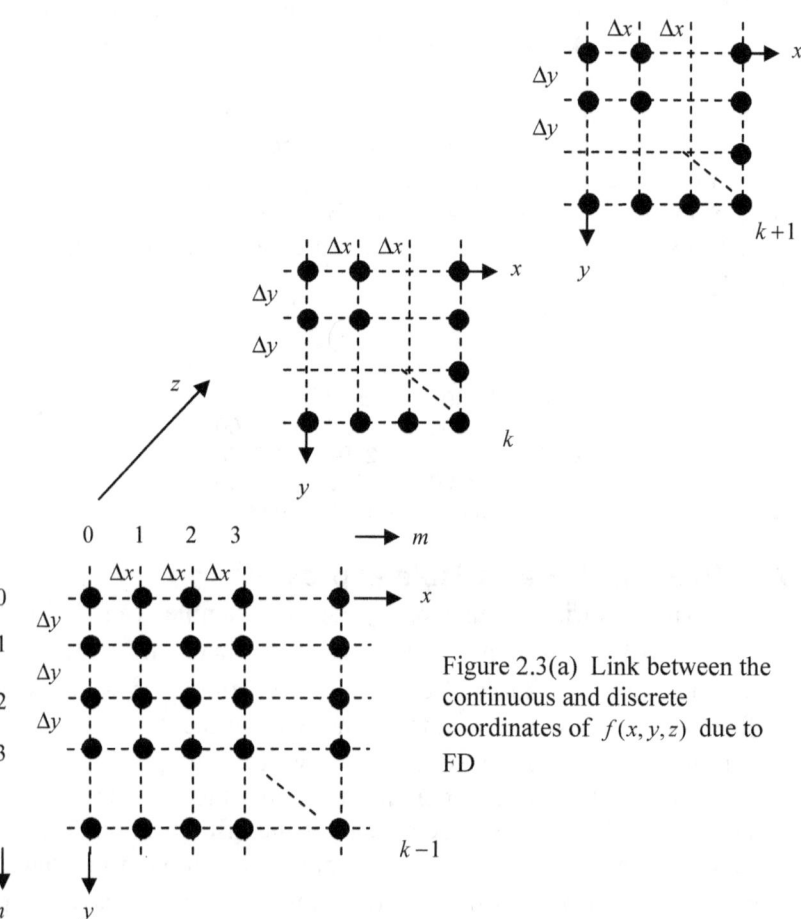

Continuous function $f(x, y, z)$

⟹

Discrete function $f[m, n, k]$

over $0 \le x \le x_0$,

$0 \le y \le y_0$, and

$0 \le z \le z_0$ with range

$f_{\min} \le f(x, y, z) \le f_{\max}$

Subject to $x = m\Delta x$,

$y = n\Delta y$, and

$z = k\Delta z$, and

$\hat{f}(x, y, z) = f_{\min} + \Delta f \; f[m, n, k]$

Figure 2.3(a) Link between the continuous and discrete coordinates of $f(x, y, z)$ due to FD

What changes do we perceive for the quantities of section 2.1? In figure 2.1(b) the vertical axis should be for $f(x, y, z)$ maintaining the variation from $f_{\min}$ to $f_{\max}$. The $f_{\min}$ is now for all pages of figure 2.3(a), so is the $f_{\max}$.

The horizontal axis of figure 2.1(b) is in x, y, z, or all three directions. The discrete form of $f(x,y,z)$ is now $f[m,n,k]$ while the reconstructed counterpart is $\hat{f}(x,y,z)$ computed from $\hat{f}(x,y,z) = f_{min} + \Delta f \; f[m,n,k]$. The level number L is taken from all pages of figure 2.3(a). The z should also have an interval which is $0 < z \le z_0$. Discretization summary context of the three dimensional counterpart is depicted in figure 2.3(a) too.

2.5 3D array for 3D finite difference

Similar to section 2.2, $f(x,y,z)$ has also matrix link but in three dimensional sense. We address three dimensional array context of deriving discrete $f[m,n,k]$ from continuous $f(x,y,z)$ by choosing a numerical example in this section.

To see the computing link, let us say $f(x,y,z) = (x-y+2z)^2$. The x-y-z domain and its sample periods are also required so let them be $-0.5 \le x \le 0.5$, $-1 \le y \le 1$, $0 \le z \le 2$, $\Delta x = 0.5$, $\Delta y = 1$, and $\Delta z = 1$.

Concerning section 2.4, we freeze every z sample and consider the whole 2D domain stretched by $-0.5 \le x \le 0.5$ and $-1 \le y \le 1$. For the given sampling periods, the x and y sample points both as a row matrix are $[-0.5 \; 0 \; 0.5]$ and $[-1 \; 0 \; 1]$ respectively. The third variable z sample is $[0 \; 1 \; 2]$. Sample values of $f(x,y,z)$ are organized as follows:

$$
\begin{array}{l}
x \rightarrow \\
y \\
\downarrow
\end{array}
\begin{bmatrix}
f(-0.5,-1,0) & f(0,-1,0) & f(0.5,-1,0) \\
f(-0.5,0,0) & f(0,0,0) & f(0.5,0,0) \\
f(-0.5,1,0) & f(0,1,0) & f(0.5,1,0)
\end{bmatrix} \; for \; z = 0,
$$

$$
\begin{array}{l}
x \rightarrow \\
y \\
\downarrow
\end{array}
\begin{bmatrix}
f(-0.5,-1,1) & f(0,-1,1) & f(0.5,-1,1) \\
f(-0.5,0,1) & f(0,0,1) & f(0.5,0,1) \\
f(-0.5,1,1) & f(0,1,1) & f(0.5,1,1)
\end{bmatrix} \; for \; z = 1, \; and
$$

$$
\begin{array}{l}
x \rightarrow \\
y \\
\downarrow
\end{array}
\begin{bmatrix}
f(-0.5,-1,2) & f(0,-1,2) & f(0.5,-1,2) \\
f(-0.5,0,2) & f(0,0,2) & f(0.5,0,2) \\
f(-0.5,1,2) & f(0,1,2) & f(0.5,1,2)
\end{bmatrix} \; for \; z = 2.
$$

Similar to section 2.2, we coalesced all x only grid points for all z's which take the following three dimensional pattern (call it X):

$$X =$$

$$\begin{bmatrix} -0.5 & 0 & 0.5 \\ -0.5 & 0 & 0.5 \\ -0.5 & 0 & 0.5 \end{bmatrix}$$
for $z = 0$

$$\begin{bmatrix} -0.5 & 0 & 0.5 \\ -0.5 & 0 & 0.5 \\ -0.5 & 0 & 0.5 \end{bmatrix}$$
for $z = 1$

$$\begin{bmatrix} -0.5 & 0 & 0.5 \\ -0.5 & 0 & 0.5 \\ -0.5 & 0 & 0.5 \end{bmatrix}$$
for $z = 2$

Similarly for all y only grid points, we have the following three dimensional array (call it Y):

$$Y =$$

$$\begin{bmatrix} -1 & -1 & -1 \\ 0 & 0 & 0 \\ 1 & 1 & 1 \end{bmatrix}$$
for $z = 0$

$$\begin{bmatrix} -1 & -1 & -1 \\ 0 & 0 & 0 \\ 1 & 1 & 1 \end{bmatrix}$$
for $z = 1$

$$\begin{bmatrix} -1 & -1 & -1 \\ 0 & 0 & 0 \\ 1 & 1 & 1 \end{bmatrix}$$
for $z = 2$

Nevertheless z only grid points are associated as well which are (call it Z):

$$Z =$$

$$\begin{bmatrix} 0 & 0 & 0 \\ 0 & 0 & 0 \\ 0 & 0 & 0 \end{bmatrix}$$
for $z = 0$

$$\begin{bmatrix} 1 & 1 & 1 \\ 1 & 1 & 1 \\ 1 & 1 & 1 \end{bmatrix}$$
for $z = 1$

$$\begin{bmatrix} 2 & 2 & 2 \\ 2 & 2 & 2 \\ 2 & 2 & 2 \end{bmatrix}$$
for $z = 2$

Having found the three three dimensional grid point arrays, the samples of $f(x,y,z)$ are computed by $f(X,Y,Z)=(X-Y+2Z)^2 =$

$$\begin{bmatrix} 0.25 & 1 & 2.25 \\ 0.25 & 0 & 0.25 \\ 2.25 & 1 & 0.25 \end{bmatrix}, \begin{bmatrix} 6.25 & 9 & 12.25 \\ 2.25 & 4 & 6.25 \\ 0.25 & 1 & 2.25 \end{bmatrix}, \text{ and } \begin{bmatrix} 20.25 & 25 & 30.25 \\ 12.25 & 16 & 20.25 \\ 6.25 & 9 & 12.25 \end{bmatrix}.$$
for $z = 0$ ..... for $z = 1$ ..... for $z = 2$

An algorithm is essential to obtain the samples of $f(x,y,z)$ in FD domain resembling section 2.2:

(a) generate x samples as a row matrix from given x variation and Δx,

(b) generate y samples as a column matrix from given y variation and Δy,

(c) generate z samples as a row or column matrix from given z variation and Δz,

(d) repeat x samples according to the number of y samples to get X but for every z sample where X is a dimensional array,

(e) repeat y samples according to the number of x samples to get Y but for every z sample where Y is a dimensional array,

(f) repeat every z sample according to the dimensions of x and y samples to get Z where Z is a dimensional array,

(g) make sure three dimensional dimensions of X, Y, and Z are the same, and

(h) finally conduct $f(X,Y,Z)$ operation in order to attain the samples of $f(x,y,z)$.

✦ How to get $f[m,n,k]$ from sampled $f(x,y,z)$?

Any three dimensional array just represents samples of $f(x,y,z)$ without Δx, Δy, or Δz information. From the three dimensional array samples we determine the following:

f_{min} =minimum in the whole $f(x,y,z)$ array and

f_{max} =maximum in the whole $f(x,y,z)$ array.

For the earlier numerical $f(x,y,z)$ we have $f_{min}=0$ and $f_{max}=30.25$ from the whole 3D array. The functional resolution Δf requires L say $L=6$ so $\Delta f =(30.25-0)/(6-1)=6.05$.

In order to get $f[m,n,k]$, rounding of $\dfrac{f(X,Y,Z)-f_{min}}{\Delta f}$ towards nearest integer is performed so earlier computed $f(X,Y,Z)$ provides $f[m,n,k]=$

$$\text{round}\{\dfrac{f(X,Y,Z)-0}{6.05}\}=\begin{bmatrix} 0 & 0 & 0 \\ 0 & 0 & 0 \\ 0 & 0 & 0 \end{bmatrix} \text{ for } k=0, \begin{bmatrix} 1 & 1 & 2 \\ 0 & 1 & 1 \\ 0 & 0 & 0 \end{bmatrix} \text{ for } k=1, \text{ and}$$

$$\begin{bmatrix} 3 & 4 & 5 \\ 2 & 3 & 3 \\ 1 & 1 & 2 \end{bmatrix} \text{ for } k=2 \text{ (also } f[m,n,0], f[m,n,1], \text{ and } f[m,n,2] \text{ respectively)}$$

thereby evolving a three dimensional array like figure 2.3(a).

✦ How to get reconstructed $\hat{f}(x,y,z)$ from sampled $f[m,n,k]$?

The reconstructed $\hat{f}(x,y,z)$ is $f_{min}+\Delta f\ f[m,n,k]$ which becomes

$$\hat{f}(x,y,z)=\begin{bmatrix} 0 & 0 & 0 \\ 0 & 0 & 0 \\ 0 & 0 & 0 \end{bmatrix} \text{ for } z=0, \begin{bmatrix} 6.05 & 6.05 & 12.1 \\ 0 & 6.05 & 6.05 \\ 0 & 0 & 0 \end{bmatrix} \text{ for } z=1, \text{ and}$$

$$\begin{bmatrix} 18.15 & 24.2 & 30.25 \\ 12.1 & 18.15 & 18.15 \\ 6.05 & 6.05 & 12.1 \end{bmatrix} \text{ for } z=2 \text{ (also } \hat{f}(x,y,0), \hat{f}(x,y,1), \text{ and } \hat{f}(x,y,2)$$

respectively) and evolves as a three dimensional array like figure 2.3(a).

❖ How to calculate the mean square error?

Now the mean square error (mse) is defined as $\dfrac{1}{M \times N \times O}\sum_z \sum_y \sum_x [f(x,y,z) - \hat{f}(x,y,z)]^2$ over the sample space where M, N, and O are the number of columns, the number of rows, and the number of two dimensional or x-y pages in $f(x,y,z)$ samples respectively, which is the average of all elements on three dimensional elements of $[f(x,y,z) - \hat{f}(x,y,z)]^2$.

For the ongoing example we obtain the error $f(x,y,z) - \hat{f}(x,y,z)$ samples as $\begin{bmatrix} 0.25 & 1 & 2.25 \\ 0.25 & 0 & 0.25 \\ 2.25 & 1 & 0.25 \end{bmatrix}$ for $z=0$, $\begin{bmatrix} 0.2 & 2.95 & 0.15 \\ 2.25 & -2.05 & 0.2 \\ 0.25 & 1 & 2.25 \end{bmatrix}$ for $z=1$, and $\begin{bmatrix} 2.1 & 0.8 & 0 \\ 0.15 & -2.15 & 2.1 \\ 0.2 & 2.95 & 0.15 \end{bmatrix}$ for $z=2$ (forms a three dimensional array too similar to other samples) and the computation is mse=$\{(0.25^2 + 1^2 + \ldots) + (0.2^2 + 2.95^2 + \ldots) + (2.1^2 + 0.8^2 + \ldots)\}/27 = 2.2015$ where $M=3$, $N=3$, and $O=3$.

❖ How to get alike convention on coordinates of 3D array?

Figures 2.3(b) and 2.3(c) correspond to the coordinate conventions of textbook and MATLAB respectively. In two dimensional counterpart we flipped the y directed data but that is not the case for 3D coordinate. If you notice carefully, basically the two are identical coordinate system just arranged differently. Although the two coordinates are identical, there is a difference regarding the origin or reference point.

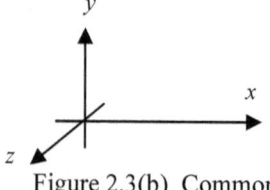

Figure 2.3(b) Common coordinate convention

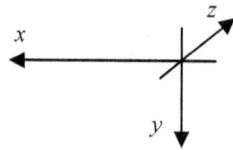

Figure 2.3(c) MATLAB coordinate convention

2.6 Implementing three dimensional FD

Our text book page, computer monitor screen, most study materials, etc are two dimensional, in some degree our visual perception too. Explained $\{x,y,z,f(x,y,z)\}$ or $\{m,n,k,f[m,n,k]\}$ are a single unit for every three dimensional expression or data that makes difficult to represent it in a page which is why a three dimensional array (appendix B.7) is necessary to exercise the finite difference. Section 2.3 is the prerequisite for this section. The samples for $\{x,y,z\}$ or $\{m,n,k\}$ refer to integer position index of the

3D array whereas value stored at some position of the array indicates functional value on $f(x,y,z)$ or $f[m,n,k]$.

Presented numerical example of the last section will be implemented here in MATLAB. The procedure as quoted is exercised in conjunction with the meshgrid similar to two dimensional counterpart.

This time meshgrid returns three grid point arrays (i.e. X, Y, and Z) each as a three dimensional array with the syntax [user-supplied variable for X, user-supplied variable for Y, user-supplied variable for Z]= meshgrid(x sample points as a row matrix, y sample points as a row matrix, z sample points as a row matrix) that is:

```
>>x=-0.5:0.5:0.5; ↵          ← x samples are assigned to x, x is user-chosen
>>y=-1:1:1; ↵                ← y samples are assigned to y, y is user-chosen
>>z=0:1:2; ↵                 ← z samples are assigned to z, z is user-chosen
>>[X,Y,Z]=meshgrid(x,y,z) ↵       ← Generation of grid point arrays i.e.
                                       X⇔X , Y⇔Y , Z⇔Z
```

```
X(:,:,1) =
   -0.5000    0   0.5000
   -0.5000    0   0.5000
   -0.5000    0   0.5000
X(:,:,2) =
   -0.5000    0   0.5000
   -0.5000    0   0.5000
   -0.5000    0   0.5000
X(:,:,3) =
   -0.5000    0   0.5000
   -0.5000    0   0.5000
   -0.5000    0   0.5000
Y(:,:,1) =
   -1   -1   -1
    0    0    0
    1    1    1
Y(:,:,2) =
   -1   -1   -1
    0    0    0
    1    1    1
Y(:,:,3) =
   -1   -1   -1
    0    0    0
    1    1    1
Z(:,:,1) =
    0    0    0
    0    0    0
    0    0    0
Z(:,:,2) =
    1    1    1
    1    1    1
    1    1    1
Z(:,:,3) =
    2    2    2
    2    2    2
    2    2    2
```

After obtaining the X, Y, and Z arrays, the $f(X,Y,Z)$ is calculated by scalar code of appendix A but as a three dimensional array so uncomplicatedly just call for the $f(x,y,z)$ samples:

```
>>f=(X-Y-2*Z).^2 ↵
```

f(:,:,1) =

0.2500	1.0000	2.2500
0.2500	0	0.2500
2.2500	1.0000	0.2500

f(:,:,2) =

2.2500	1.0000	0.2500
6.2500	4.0000	2.2500
12.2500	9.0000	6.2500

f(:,:,3) =

12.2500	9.0000	6.2500
20.2500	16.0000	12.2500
30.2500	25.0000	20.2500

As the return exposes, f(:,:,1), f(:,:,2), and f(:,:,3) refer to the samples of $f(x,y,z)$ for $z=0$, 1, and 2 equivalently $f(x,y,0)$, $f(x,y,1)$, and $f(x,y,2)$ respectively.

Consequentially the three dimensional discrete array $f[m,n,k]$ needs $f(x,y,z)|_{min}$ and $f(x,y,z)|_{max}$ in the sample space. Now the hurdle is min or max works for a row or column matrix so we turn the 3D array as a single column by f(:) and exercise min or max over f(:) as follows:

```
>>f1=min(f(:)); ↵          ← f1⇔ f(x,y,z)|min, f1 is user-chosen
>>f2=max(f(:)); ↵          ← f2⇔ f(x,y,z)|max, f2 is user-chosen
```

In addition level number L and functional resolution Δf are associated too:

```
>>L=6; df=(f2-f1)/(L-1); ↵   ← L⇔ L , df⇔ Δf , L and df are user-chosen
```

Thus computing for $f[m,n,k]$ as a three dimensional array requires:

```
>>fmnk=round((f-f1)/df) ↵   ← fmnk⇔ f[m,n,k], fmnk is user-chosen
```

fmnk(:,:,1) =

0	0	0
0	0	0
0	0	0

fmnk(:,:,2) =

0	0	0
1	1	0
2	1	1

fmnk(:,:,3) =

2	1	1
3	3	2
5	4	3

After that the reconstructed $\hat{f}(x,y,z)$ samples as a three dimensional array we obtain by $f_{min}+\Delta f\ f[m,n,k]$:

```
>>f_hat=f1+df*fmnk ↵        ← f_hat⇔ f̂(x,y,z) , f_hat is user-chosen
```

f_hat(:,:,1) =

0	0	0
0	0	0

$$0 \quad 0 \quad 0$$

f_hat(:,:,2) =

0	0	0
6.0500	6.0500	0
12.1000	6.0500	6.0500

f_hat(:,:,3) =

12.1000	6.0500	6.0500
18.1500	18.1500	12.1000
30.2500	24.2000	18.1500

The error samples due to FD are $f(x,y,z) - \hat{f}(x,y,z)$ and obtain them as a three dimensional array by:

>>e=f-f_hat ↵   ← e⇔ $f(x,y,z) - \hat{f}(x,y,z)$, e is user-chosen

e(:,:,1) =

0.2500	1.0000	2.2500
0.2500	0	0.2500
2.2500	1.0000	0.2500

e(:,:,2) =

2.2500	1.0000	0.2500
0.2000	-2.0500	2.2500
0.1500	2.9500	0.2000

e(:,:,3) =

0.1500	2.9500	0.2000
2.1000	-2.1500	0.1500
0	0.8000	2.1000

Lastly the mean square error is to be computed but the error samples are in a three dimensional array that is why first convert it to a column by e(:) and then call the error finder:

>>mse(e(:)) ↵

ans =

2.2015

✦ How to get conventional arrangement of samples?

After computing whatever three dimensional array we have obtained from MATLAB follows the coordinate convention of figure 2.3(c). What if we wish to obtain the conventional data arrangement like figure 2.3(b)? We need to flip every z directed planar data both in x and y directions and which happens by fliplr and flipud respectively (appendix B.6). We may use a for-loop (appendix B.4) to have control on every planar data which is demonstrated for the $f(x,y,z)$ samples (stored in f) in the following:

```
for k=1:3
    f(:,:,k)=fliplr(flipud(f(:,:,k)));
end
```

If you call f, you will see the last section arranged samples of $f(x,y,z)$. This tactic can be exercised on other samples i.e. on e, f_hat, etc.

2.7 Discrete 2D gradient or differentiation

In order to determine the 2D gradient we start from integer $f[m,n]$ or $f(x,y)$ samples. In either function we acquire a rectangular matrix of coarse the resolution information must be known (i.e. Δx and Δy).

Given information may appear in two forms - values of $f[m,n]$ or $f(x,y)$ as a rectangular matrix and expression of $f(x,y)$. Say any FD point p has the coordinate (m,n), the surrounding grid points have the coordinates as shown below:

●	●	●	●	●
●	● $(m-1,n-1)$	● $(m,n-1)$	● $(m+1,n-1)$	●
●	● $(m-1,n)$	p ● (m,n)	● $(m+1,n)$	●
●	● $(m-1,n+1)$	● $(m,n+1)$	● $(m+1,n+1)$	●
●	●	●	●	●

● represents one grid point

We assume that the reader is familiar with the basic difference definitions [30]. There are four difference definitions (divided, forward, backward, and central), one of which is the forward difference of the first order i.e.

m directed gradient: $G_m = \dfrac{\Delta f[m,n]}{\Delta m} = f[m+1,n] - f[m,n]$,

n directed gradient: $G_n = \dfrac{\Delta f[m,n]}{\Delta n} = f[m,n+1] - f[m,n]$, and

magnitude gradient: $G = \sqrt{G_m^2 + G_n^2}$.

The two gradients are actually linked to its continuous counterparts $\dfrac{\partial f(x,y)}{\partial x}$ and $\dfrac{\partial f(x,y)}{\partial y}$ respectively.

If second order gradient is sought, the expressions we need are the following:

m directed gradient: $G_{2m} = f[m+2,n] - 2f[m+1,n] + f[m,n]$,

n directed gradient: $G_{2n} = f[m,n+2] - 2f[m,n+1] + f[m,n]$, and

magnitude gradient: $G_2 = \sqrt{G_{2m}^2 + G_{2n}^2}$.

In continuous counterpart of the first order derivative, the $\dfrac{\partial f(x,y)}{\partial x}$ is infinite if the $f(x,y)$ changes abruptly at some x for example 0 to 1 at $x=1$. But in discrete mathematics we do not handle infinity. We just get a large number at

x where the abruptness occurs. The G_m, G_n, and G can be written as $G_m[m,n]$, $G_n[m,n]$, and $G[m,n]$ with the same coordinate convention indicating another discrete function respectively. Following examples illustrate the computing on the 2D gradient.

✦ Example 1: m / n directed first order gradient on $f[m,n]$

Let us choose $f[m,n] = \begin{bmatrix} 9 & 45 & 43 & 9 \\ 4 & 32 & 45 & 6 \\ 8 & 21 & 34 & 6 \end{bmatrix}$ for the computing. There is an

embedded function called **gradient** which operates in a slight different way. At the beginning and ending elements the function computes the forward and backward differences respectively whereas the inside elements are treated based on the central difference.

The $f[m,n]$ has the first row [9 45 43 9]. Both edge elements are 9 so the m directed gradient at those positions should be 45–9 and 9–43 respectively. At the element 45, the gradient value is (43–9)/2=17. Continuing this way we get the first order m directed gradient as [36 17 –18 –34]. For the whole $f[m,n]$ we should have $G_m = \begin{bmatrix} 36 & 17 & -18 & -34 \\ 28 & 20.5 & -13 & -39 \\ 13 & 13 & -7.5 & -28 \end{bmatrix}$.

Likewise calculation along n direction yields $G_n = \begin{bmatrix} -5 & -13 & 2 & -3 \\ -0.5 & -12 & -4.5 & -1.5 \\ 4 & -11 & -11 & 0 \end{bmatrix}$.

The **gradient** needs two output arguments with the syntax $[G_m, G_n] =$ gradient($f[m,n]$) (obviously $f[m,n]$ entered as a rectangular matrix which is f) so call it as follows:

```
>>f=[9 45 43 9;4 32 45 6;8 21 34 6]; ↵
>>[Gm,Gn]=gradient(f) ↵
```

Gm =

36.0000	17.0000	-18.0000	-34.0000
28.0000	20.5000	-13.0000	-39.0000
13.0000	13.0000	-7.5000	-28.0000

Gn =

-5.0000	-13.0000	2.0000	-3.0000
-0.5000	-12.0000	-4.5000	-1.5000
4.0000	-11.0000	-11.0000	0

✦ Example 2: Gradient from functional samples

In section 2.3 functional samples for $f(x,y)$ are available in the variable f. In example 1 we did not mention any resolution information which was $\Delta x = 0.5$ and $\Delta y = 0.5$ for the $f(x,y)$ in section 2.3. The **gradient** keeps provision for resolution input too which is exercised by adding two input arguments i.e. [Gx,Gy]=gradient(f, Δx , Δy) hence the calling will be:

```
>>[Gx,Gy]=gradient(f,0.5,0.5); ↵
```

2.8 Discrete 3D gradient or differentiation

Last section explained **gradient** is so versatile that it accommodates three dimensional gradient too. In order to determine the 3D gradient we start from the integer $f[m,n,k]$ or $f(x,y,z)$ samples not to mention each being a three dimensional array subject to resolutions Δx, Δy, and Δz. If we provide resolutions, **gradient** computes divided difference else modified difference like last section. The three first order discrete gradients or differentiations are as follows:

m directed gradient: $G_m = \dfrac{\Delta f[m,n,k]}{\Delta m} = f[m+1,n,k] - f[m,n,k]$,

n directed gradient: $G_n = \dfrac{\Delta f[m,n,k]}{\Delta n} = f[m,n+1,k] - f[m,n,k]$,

k directed gradient: $G_k = \dfrac{\Delta f[m,n,k]}{\Delta k} = f[m,n,k+1] - f[m,n,k]$, and

magnitude gradient: $G = \sqrt{G_m^2 + G_n^2 + G_k^2}$.

The symbols G_m, G_n, and G_k correspond to $\dfrac{\partial f(x,y,z)}{\partial x}$, $\dfrac{\partial f(x,y,z)}{\partial y}$, and $\dfrac{\partial f(x,y,z)}{\partial z}$ respectively. All other symbols have previous section quoted meanings. Each of the Δm, Δn, and Δk is 1. We wish to compute the discrete gradient of section 2.6 quoted three dimensional function.

```
>>x=-0.5:0.5:0.5; y=-1:1:1; z=0:1:2; [X,Y,Z]=meshgrid(x,y,z); f=(X-Y-2*Z).^2; ↵
>>f1=min(f(:)); f2=max(f(:)); L=6; df=(f2-f1)/(L-1); fmnk=round((f-f1)/df); ↵
```

Last two command lines are the repetition of section 2.6 exercised commands manifesting the availability of $f(x,y,z)$ samples in **f** and $f[m,n,k]$ samples in **fmnk**.

The syntax for the three dimensional gradient is $[G_m,G_n,G_k] =$ **gradient**($f[m,n,k]$ samples) and $[G_x,G_y,G_z] =$ **gradient**($f(x,y,z)$ samples, Δx, Δy, Δz) for the modified difference and its divided counterpart respectively. It is important to point out that the returned gradient G_m or G_x is a three dimensional array, so are the other two components.

For the $f[m,n,k]$ gradient, execute the following:
```
>>[Gm,Gn,Gk]=gradient(fmnk); ↵
```

For the $f(x,y,z)$ gradient, execute the following:
```
>>[Gx,Gy,Gz]=gradient(f,0.5,1,1); ↵
```
In case magnitude gradient is sought, following is carried out:
```
>>G=sqrt(Gx.^2+Gy.^2+Gz.^2); ↵
```

Every element in above return i.e. **Gm, Gx, G,** etc is a three dimensional array with earlier mentioned coordinate system.

◆ Theoretical linkage of 3D discrete gradient

Although so explained **gradient** keeps provision for returning the 2D or 3D discrete differentiations, its definition is in mixed form rather theoretical context. Some problems need theoretical definitions which we address now and the discourse is on first 2D then 3D.

There is another embedded function by the name **diff** which provides forward or backward difference as required. In order to understand its execution we consider the matrix $f[m,n]$ of (example 1) section 2.7. Enter the matrix to **f** and exercise the **diff** over **f**:

```
>>f=[9 45 43 9;4 32 45 6;8 21 34 6]; ↵
>>diff(f) ↵
```

```
ans =
    -5  -13    2   -3
     4  -11  -11    0
```

The **diff(f)** basically returns $G_n = \dfrac{\Delta f[m,n]}{\Delta n} = f[m,n+1] - f[m,n]$ (forward difference). We know from the theory of difference that the number of samples for G_n is one less than that of the $f[m,n]$ in the direction of n or columns. That is why G_n has two rows although $f[m,n]$ does 3. How to determine then G_m? You need to transpose $f[m,n]$, exercise **diff** over the transposed $f[m,n]$, and transpose (' is the transpose operator in MATLAB) the result that is:

```
>>Gm=diff(f')' ↵
```

```
Gm =
    36   -2  -34
    28   13  -39
    13   13  -28
```

What does above **Gm** indicate?, certainly $G_m = \dfrac{\Delta f[m,n]}{\Delta m} = f[m+1,n] - f[m,n]$. In the m or row direction we see 3 samples while $f[m,n]$ has 4 (for the reason stated for G_n).

What about the divided difference? We need the samples of $f(x,y)$. Consider the example 1 of section 2.3 where **f** holds the samples of $f(x,y)$. Execute the following:

```
>>Gx=diff(f')'/0.5; Gy=diff(f)/0.5; ↵
```

Above **Gx** and **Gy** retain the samples of $G_x = \dfrac{\partial f(x,y)}{\partial x} \approx \dfrac{f(x+\Delta x, y) - f(x,y)}{\Delta x}$

and $G_y = \dfrac{\partial f(x,y)}{\partial y} \approx \dfrac{f(x,y+\Delta y) - f(x,y)}{\Delta y}$ respectively where **Gx** and **Gy** are user-chosen and $\Delta x = 0.5$ and $\Delta y = 0.5$ for the example.

These implementations we extend to 3D with sight programming twist. The 3D example from section 2.6 is chosen and reexcute the following:

```
>>x=-0.5:0.5:0.5;y=-1:1:1;z=0:1:2;[X,Y,Z]=meshgrid(x,y,z); f=(X-Y-2*Z).^2; ↵
```
The samples of $f(x,y,z)$ are available in f. The three first order discrete forward derivatives are defined as follows:

$$G_x = \frac{\partial f(x,y,z)}{\partial x} \approx \frac{f(x+\Delta x, y, z) - f(x,y,z)}{\Delta x},$$

$$G_y = \frac{\partial f(x,y,z)}{\partial y} \approx \frac{f(x, y+\Delta y, z) - f(x,y,z)}{\Delta y}, \text{ and}$$

$$G_z = \frac{\partial f(x,y,z)}{\partial z} \approx \frac{f(x, y, z+\Delta z) - f(x,y,z)}{\Delta z}.$$

The $f(x,y,z)$ samples are stored in f as 3D array that is why it can not be handled the way we did the 2D counterpart moreover diff does not determine all derivatives though it accepts 3D array. Following you may execute (appendix B.4 for loop) for the G_x :

```
>>for p=1:3, Gx(:,:,p)=diff(f(:,:,p)')'/0.5; end ↵
>>Gx ↵
```

```
Gx(:,:,1) =
          1.5000   2.5000
         -0.5000   0.5000
         -2.5000  -1.5000
Gx(:,:,2) =
         -2.5000  -1.5000
         -4.5000  -3.5000
         -6.5000  -5.5000
Gx(:,:,3) =
         -6.5000   -5.5000
         -8.5000   -7.5000
        -10.5000   -9.5000
```

In above along the row direction of every page in Gx you find one column reduced in accordance with the definition. There are three pages which is why maximum for loop counter is 3. The for loop counter p is user-chosen and Gx(:,:,p) indicates the p-th page in Gx which holds G_x as a 3D array. The f(:,:,p)' is the transposition of f(:,:,p) like the 2D counterpart. Actually we exercised diff on page by page basis. Not to mention the resolution $\Delta x = 0.5$ takes part in the computing too.

Anyhow the G_y does not need a for loop and get it with $\Delta y = 0.5$ by:

```
>>Gy=diff(f)/0.5 ↵
```

```
Gy(:,:,1) =
         0  -2  -4
         4   2   0
Gy(:,:,2) =
         8   6   4
        12  10   8
Gy(:,:,3) =
        16  14  12
        20  18  16
```

Owing to the functionality of **diff** and the presence of third dimension in G_z we need a for loop for the G_z :

```
>>for p=1:2, Gz(:,:,p)=(f(:,:,p+1)-f(:,:,p))/1; end ↵
>>Gz ↵

Gz(:,:,1) =
        2    0   -2
        6    4    2
       10    8    6
Gz(:,:,2) =
       10    8    6
       14   12   10
       18   16   14
```

The G_z does not have 3 pages as $f(x,y,z)$ samples do. Why so? The definition quoted reduction is now in the page direction. Page by page circumstance for the forward difference definition is exercised by using f(:,:,p+1)-f(:,:,p) i.e. $f(x,y,z+\Delta z)-f(x,y,z)$ not to mention $\Delta z=1$ for the example at hand. Despite three pages we used for loop counter as 2 which is coming from the number of pages minus one. The **Gz** holds G_z samples as a three dimensional array like the other two counterparts.

The bottom line is whichever direction derivative you compute for the samples of $f(x,y,z)$ will be one less in that direction whether x, y, or z.

As another example compute the forward discrete gradients G_x, G_y, and G_z for $f(x,y,z)=e^{-x-y-z}\cos(x-y+2z)$ over $-0.25 \le x \le 0.3$, $-2 \le y \le 3$, and $0 \le z \le 0.7$ where $\Delta x =0.05$, $\Delta y =0.5$, and $\Delta z =0.1$.

Grid point and functional computings are as follows:
```
>>x=-0.25:0.05:0.3; y=-2:0.5:3; z=0:0.1:0.7; [X,Y,Z]=meshgrid(x,y,z); ↵
>>f=exp(-X-Y-Z).*cos(X-Y+2*Z); ↵
```
It is better if we enter the resolutions to like name variables e.g. **dx** for Δx :
```
>>dx=0.05; dy=0.5; dz=0.1; ↵
```
In the case of sample number finding we may exercise **length** over x, y, or z so the G_x is computed by:
```
>>for p=1:length(z), Gx(:,:,p)=diff(f(:,:,p)')'/dx; end ↵
```
Although G_x is being computed in above, **length(z)** is used because of page by page reason. However the other two counterparts are obtained by the following:
```
>>Gy=diff(f)/dy; ↵
>>for p=1:length(z)-1, Gz(:,:,p)=(f(:,:,p+1)-f(:,:,p))/dz; end ↵
```
If you notice the 3D array sizes of $f(x,y,z)$, G_x, G_y, and G_z, they are 11×12×8, 11×11×8, 10×12×8, and 11×12×7 respectively obviously not identical for this reason earlier **gradient** is modified.

Amazingly the **diff(f,1,2)/dx**, **diff(f,1,1)/dy**, and **diff(f,1,3)/dz** would function for **Gx**, **Gy**, and **Gz** respectively, execute **help diff** for details.

2.9 Interpolation on two dimensional FD data

Starting from integer $f[m,n]$ or $f(x,y)$ samples, objective of two dimensional (2D) interpolation is to obtain intermediate samples based on some user-defined resolution or step size.

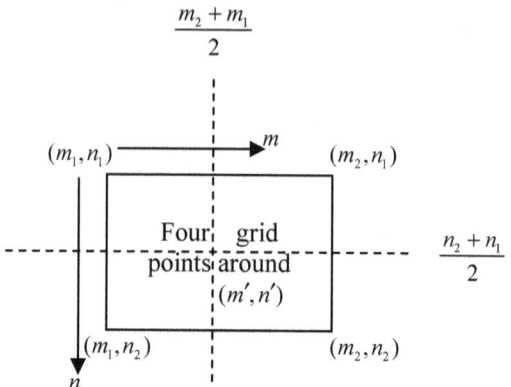

Figure 2.4(a) Squarely formed four grid
points in 2D FD domain

Grid point coordinate of a discrete function $f[m,n]$ is integer. Figure 2.4(a) depicts the (m',n') inside a square formed by four grid points in 2D FD domain. The (m',n') is assumed to be fractional. Suppose we need $f[m',n']$ which inevitably involves 2D interpolation. There is no absolute guarantee that $f[m',n']$ value obtained this way is close to the actual one however better results are achieved due to an interpolation.

We wish to address the theory of nearest neighborhood and bilinear interpolations as a basic introduction.

✦ Nearest neighborhood interpolation

The fractional (m',n') lies within the squarely formed four integer coordinates (m_1,n_1), (m_2,n_1), (m_2,n_2), and (m_1,n_2) as pointed out in figure 2.4(a). In this scheme we approximate the wanted $f[m',n']$ to the closest of the four functional values. Mathematically the approximation is carried out as follows:

$$\text{if } m' \geq \frac{m_2 + m_1}{2}, \text{ then } m' = m_2$$

$$\text{else } m' = m_1 \text{ and}$$

$$\text{if } n' \geq \frac{n_2 + n_1}{2}, \text{ then } n' = n_2$$

$$\text{else } n' = n_1.$$

Following numerical example demonstrates the nearest neighborhood interpolation.

Example:

By applying the nearest neighborhood interpolation, determine $f[m',n']$ on $f[m,n]$ (example 1 of section 2.7) at fractional $(m',n') = (0.3,1.5)$.

The m' and n' coordinates lie between 0 and 1 and 1 and 2 respectively so $m_1 = 0$, $m_2 = 1$, $n_1 = 1$, and $n_2 = 2$. Comparing the m' and n' coordinates with $\frac{m_2 + m_1}{2}$ and $\frac{n_2 + n_1}{2}$ respectively, we get the nearest neighbor coordinate as $(0,2)$ therefore $f[m',n'] = f[0,2] = 8$ (figure 2.1(a) shown convention is applied).

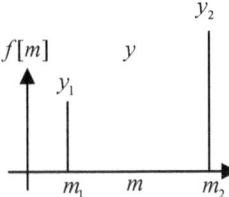

Figure 2.4(b) Two samples
of a function

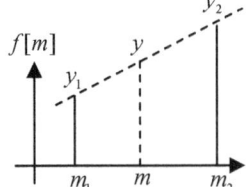

Figure 2.4(c) Linear interpolation
of a discrete function from two
sample values

♦ Bilinear interpolation

Bilinear interpolation is derived from one dimensional counterpart. Figure 2.4(b) shows the interpolation from two sample values of a linear discrete function which at this time is applied in m and n directions of $f[m,n]$ that is why the term bilinear is associated.

Concerning the figure 2.4(a), squarely formed four grid points have the coordinates and functional values (m_1,n_1), (m_2,n_1), (m_2,n_2), and (m_1,n_2) and $f[m_1,n_1]$, $f[m_2,n_1]$, $f[m_2,n_2]$, and $f[m_1,n_2]$ respectively. We determine $f[m',n']$ at fractional (m',n') from these four grid points. Following expressions are exercised to calculate the bilinearly interpolated functional value:

linearity along $n = n_1$:

$$f[m',n_1] = \frac{f[m_2,n_1] - f[m_1,n_1]}{m_2 - m_1}(m' - m_1) + f[m_1,n_1]$$

linearity along $n = n_2$:

$$f[m',n_2] = \frac{f[m_2,n_2] - f[m_1,n_2]}{m_2 - m_1}(m' - m_1) + f[m_1,n_2]$$

linearity along $m = m'$:

$$f[m',n'] = \frac{f[m',n_2] - f[m',n_1]}{n_2 - n_1}(n' - n_1) + f[m',n_1]$$

Understandably the last mathematical relationships are valid over $m_1 \le m' \le m_2$ and $n_1 \le n' \le n_2$.

Example:

Apply bilinear interpolation to determine the $f[m',n']$ at $(m',n') = (0.3,1.5)$ for example 1 of section 2.7 mentioned $f[m,n]$.

From the coordinate of interest we find the neighboring functional values as $\begin{bmatrix} 4 & 32 \\ 8 & 21 \end{bmatrix}$ in given $f[m,n]$. Related coordinates and functional values are as follows: $m_1 = 0$, $m_2 = 1$, $n_1 = 1$, $n_2 = 2$, $f[0,1] = 4$, $f[1,1] = 32$, $f[0,2] = 8$, and $f[1,2] = 21$ hence

$$f[m',n_1] = \frac{f[1,1] - f[0,1]}{1-0}(0.3-0) + f[0,1] = (32-4)(0.3) + 4 = 12.4,$$

similarly,

$$f[m',n_2] = \frac{f[1,2] - f[0,2]}{1-0}(0.3-0) + f[0,2] = 11.9, \text{ and finally}$$

$$f[m',n'] = \frac{11.9 - 12.4}{2-1}(1.5-1) + 12.4 = 12.15.$$

✦ MATLAB embedded function

The function interp2 conducts 2D interpolations which keeps a variety of options. A simple procedure is followed:

(1) From given m and n values, obtain the grid point matrices like section 2.2.

(2) Enter the given $f[m,n]$.

(3) From wanted (m',n') values, obtain the grid point matrices like step (1).

(4) Call the interp2 with five input argument syntax interp2(m directed grid point rectangular matrix, n directed grid point rectangular matrix, given $f[m,n]$ as a rectangular matrix, wanted m' directed grid point rectangular matrix, wanted n' directed grid point rectangular matrix) where the last two input arguments can be single element too. The return from the interp2 is assigned to user-chosen variable fn which holds the interpolated $f[m',n']$.

(5) The default method chosen by the interp2 is bilinear interpolation. In case other methods are required, another input argument is appended with the interp2. Interpolation methods that are available in MATLAB are the nearest neighborhood, piecewise cubic spline, and shape preserving piecewise cubic whose MATLAB indicatory reserve words are nearest, spline, and cubic respectively, each of which is put under a quote.

✦ Interpolation at a single point by interp2

We demonstrated two numerical examples earlier considering the $f[m,n]$ of section 2.7. Let us enter the $f[m,n]$ to f as follows:

```
>>f=[9 45 43 9;4 32 45 6;8 21 34 6]; ↵
```

The matrix size of $f[m,n]$ says that the m and n intervals are $0 \le m \le 3$ and $0 \le n \le 2$ respectively upon which both variables as a row matrix are generated by:

```
>>m=0:3; ↵        ← m is user-chosen, holds all m s
>>n=0:2; ↵        ← n is user-chosen, holds all n s
```

Using the **meshgrid** of section 2.3, grid point matrices are generated by:

```
>>[M,N]=meshgrid(m,n); ↵
```

In above execution the M and N are user-chosen variables which hold the m and n directed grid point matrices respectively. Earlier we obtained $f[m',n']$ as 8 and 12.15 attributed to the nearest neighborhood and bilinear interpolations at $(m',n') = (0.3,1.5)$ respectively which we get by:

For the nearest neighborhood interpolation:
```
>>fn=interp2(M,N,f,0.3,1.5,'nearest') ↵
```

```
fn =
        8
```

For the bilinear interpolation:
```
>>fn=interp2(M,N,f,0.3,1.5) ↵
```

```
fn =
       12.1500
```

✦ Interpolation at multiple points by interp2

Suppose for the ongoing $f[m,n]$, three $f[m',n']$ s employing bilinear interpolation are 12.15, 25.72, and 26 at $(m',n') = (0.3,1.5)$, $(0.7,0.8)$, and $(1.3,1.9)$ respectively which we intend to find.

Following is the execution:
```
>>mp=[0.3 0.7 1.3]; ↵  ← Given m′ s are assigned to mp as a row matrix, where
                            mp is user-chosen
>>np=[1.5 0.8 1.9]; ↵  ← Given n′ s are assigned to np as a row matrix, where np
                            is user-chosen
>>fn=interp2(M,N,f,mp,np) ↵        ← Calling the function on points of interest
```

```
fn =
       12.1500   25.7200   26.0000
```

The return is also as a row matrix. Should the reader need each interpolated value, the command fn(1), fn(2), or fn(3) is exercised respectively. In case the nearest neighborhood method were sought, the command would be fn=interp2(M,N,f,mp,np,'nearest');.

Interpolation along a row or column can also be handled this way for instance $f[m',n']$ over $1 \le n' \le 1.5$ with $\Delta n' = 0.125$ and $m' = 0.7$ (assume bilinear interpolation) needs us to execute the following:

```
>>mp=0.7; ↵
>>np=1:0.125:1.5; ↵
>>fn=interp2(M,N,f,mp,np) ↵
```

fn =

```
            23.6000
            22.7875
            21.9750
            21.1625
            20.3500
```

The n' is in column direction that is why the return is as a column matrix.

◆ Interpolation over a domain by interp2

Instead of point set, an area can also be of interest. For the ongoing $f[m,n]$, the $f[m',n']$ is

$$\begin{bmatrix} 4 & 6.8 & 9.6 & 12.4 & 15.2 & 18 & 20.8 \\ 4.5 & 7.1125 & 9.725 & 12.3375 & 14.95 & 17.5625 & 20.175 \\ 5 & 7.425 & 9.85 & 12.275 & 14.7 & 17.125 & 19.55 \\ 5.5 & 7.7375 & 9.975 & 12.2125 & 14.45 & 16.6875 & 18.925 \\ 6 & 8.05 & 10.1 & 12.15 & 14.2 & 16.25 & 18.3 \end{bmatrix}$$

over $0 \le m' \le 0.6$ and $1 \le n' \le 1.5$ with $\Delta m' = 0.1$ and $\Delta n' = 0.125$ by using bilinear interpolation which we wish to find.

When interpolation over a domain is required, it is mandatory that we generate the rectangular grid point matrices by employing the **meshgrid** therefore execute the following:

```
>>mp=0:0.1:0.6; ↵    ← Given m's are assigned to mp as a row matrix, where
                          mp is user-chosen
>>np=1:0.125:1.5; ↵  ← Given n's are assigned to np as a row matrix, where np
                          is user-chosen
>>[Mp,Np]=meshgrid(mp,np); ↵  ← Mp and Np hold m' and n' directed grid
                          point matrices respectively where Mp and Np are user-chosen
```

Eventually call the interpolator at points of interest:

```
>>fn=interp2(M,N,f,Mp,Np) ↵        ← fn holds f[m',n'], fn is user-chosen
```

fn =

```
    4.0000   6.8000   9.6000  12.4000  15.2000  18.0000  20.8000
    4.5000   7.1125   9.7250  12.3375  14.9500  17.5625  20.1750
    5.0000   7.4250   9.8500  12.2750  14.7000  17.1250  19.5500
    5.5000   7.7375   9.9750  12.2125  14.4500  16.6875  18.9250
    6.0000   8.0500  10.1000  12.1500  14.2000  16.2500  18.3000
```

◆ Note on the interpolator

(1) When any (m',n') falls out of the given domain, the return is an undesirable character which we call not a number (NaN).

(2) Multiplicity of $\Delta m'$ or $\Delta n'$ should be maintained while exercising the range data for example on $\Delta m' = 0.4$ the interval $0 \le m' \le 3$ can not be appropriately covered whereas $\Delta m' = 0.3$ can.

(3) Whatever computing we performed on $f[m,n]$ can be conducted on the samples of $f(x,y)$.

2.10 Interpolation on three dimensional FD data

Section 2.9 is the prerequisite for this section. The three dimensional interpolation is the extension of 2D counterpart which happens by the analogous function interp3. Now our intent is to determine $f[m',n',k']$ from $f[m,n,k]$ at some point (m',n',k'). The algorithm of the last section for interp2 is equally applicable here but the third dimension is to be added.

♦ Syntax of the 3D interpolator

The syntax we need is fn=interp3(m directed grid point 3D array, n directed grid point 3D array, k directed grid point 3D array, given $f[m,n,k]$ as a 3D array, wanted m' directed grid point 3D array, wanted n' directed grid point 3D array, wanted k' directed grid point 3D array) where the last three input arguments can be single element too. The return from the interp3 is assigned to user-chosen variable fn which holds the interpolated $f[m',n',k']$ as a 3D array.

Although we explained the syntax considering $f[m',n',k']$, the interp3 functions for the $f(x',y',z')$ too.

Interpolation method you may also choose by adding another input argument at the end of interp3 like the 2D counterpart of last section.

♦ Example on a 3D interpolation

Section 2.5 illustrated function $f(x,y,z) = (x-y+2z)^2$ is implemented subject to $\Delta x = 0.5$, $\Delta y = 1$, and $\Delta z = 1$ over $-0.5 \le x \le 0.5$, $-1 \le y \le 1$, and $0 \le z \le 2$ in section 2.6. Just repeat the section 2.6 commands for f:
```
>>x=-0.5:0.5:0.5; y=-1:1:1; z=0:1:2; [X,Y,Z]=meshgrid(x,y,z); f=(X-Y-2*Z).^2; ↵
```

We know that $f(x,y,z)$ samples are available in f as a 3D array. Suppose we wish to obtain bilinear interpolated samples subject to $\Delta x = 0.25$, $\Delta y = 0.5$, and $\Delta z = 0.5$ starting from f over the same interval. As far as procedure is concerned, any interpolation point is (x',y',z') and x', y', and z' directed variations as a row matrices are generated by (xp, yp, and zp respectively):
```
>>xp=-0.5:0.25:0.5; yp=-1:0.5:1; zp=0:0.5:2; ↵
```

After that generate the grid point 3D arrays or (X',Y',Z') points by (Xp, Yp, and Zp respectively):
```
>>[Xp,Yp,Zp]=meshgrid(xp,yp,zp); ↵
```

Then call the 3D interpolator for the $f(x',y',z')$ samples:

```
>>fn=interp3(X,Y,Z,f,Xp,Yp,Zp) ↵    ← fn holds f(x',y',z'), fn is user-chosen
fn(:,:,1) =
```

0.2500	0.6250	1.0000	1.6250	2.2500
0.2500	0.3750	0.5000	0.8750	1.2500
0.2500	0.1250	0	0.1250	0.2500
1.2500	0.8750	0.5000	0.3750	0.2500
2.2500	1.6250	1.0000	0.6250	0.2500

```
fn(:,:,2) =
```

1.2500	1.1250	1.0000	1.1250	1.2500
2.2500	1.8750	1.5000	1.3750	1.2500
3.2500	2.6250	2.0000	1.6250	1.2500
5.2500	4.3750	3.5000	2.8750	2.2500
7.2500	6.1250	5.0000	4.1250	3.2500

```
fn(:,:,3) =
```

2.2500	1.6250	1.0000	0.6250	0.2500
4.2500	3.3750	2.5000	1.8750	1.2500
6.2500	5.1250	4.0000	3.1250	2.2500
9.2500	7.8750	6.5000	5.3750	4.2500
12.2500	10.6250	9.0000	7.6250	6.2500

```
fn(:,:,4) =
```

7.2500	6.1250	5.0000	4.1250	3.2500
10.2500	8.8750	7.5000	6.3750	5.2500
13.2500	11.6250	10.0000	8.6250	7.2500
17.2500	15.3750	13.5000	11.8750	10.2500
21.2500	19.1250	17.0000	15.1250	13.2500

```
fn(:,:,5) =
```

12.2500	10.6250	9.0000	7.6250	6.2500
16.2500	14.3750	12.5000	10.8750	9.2500
20.2500	18.1250	16.0000	14.1250	12.2500
25.2500	22.8750	20.5000	18.3750	16.2500
30.2500	27.6250	25.0000	22.6250	20.2500

Now the fn(:,:,1), fn(:,:,2), fn(:,:,3), etc refer to $z'=0$, $z'=0.5$, $z'=1$, etc respectively which is why 5 pages are in fn unlike 3 in f. What about the x' and y' variations? In every z directed page five columns are there which correspond to $x'=-0.5$, $x'=-0.25$, etc respectively. Again in the row direction considered new y' points are $y'=-1$, $y'=-0.5$, etc respectively certainly for every page.

If we wish to interpolate $f[m,n,k]$, the command would be (with earlier section symbol meanings):

```
>>f1=min(f(:)); f2=max(f(:)); L=6; df=(f2-f1)/(L-1); fmnk=round((f-f1)/df); ↵
>>fn=interp3(X,Y,Z,fmnk,Xp,Yp,Zp); ↵ ← fn holds f[m',n',k'], fn is user-chosen
```

It is immaterial whether we choose integer (m,n,k) or fractional (x,y,z) for the computing, either selection provides the same return. Also you may choose integer $f[m',n',k']$ or fractional $f(x',y',z')$ for the interp3.

✦ Example on a single point 3D interpolation

We illustrated the 3D interpolation of $f(x',y',z')$ over the whole (x,y,z) space. The interpolator returns a single point interpolated value too.

In that case there is no need to generate (X', Y', Z') space. Despite integer (m, n, k) the $f[m', n', k']$ is required at $(m', n', k') = (0.4, 0.4, 1.5)$ which is carried out by:

```
>>fn=interp3(X,Y,Z,fmnk,0.4,0.4,1.5) ↵

fn =
      1.5600
```

According to the last return we get $f[m', n', k'] = 1.56$.

✦ Example on specific method interpolation

In case we need other interpolation than bilinear, an input argument at the end of **interp3** like the 2D counterpart implements that. Say in the last computing the interpolation method is the nearest neighborhood, the command is then:

```
>>fn=interp3(X,Y,Z,fmnk,0.4,0.4,1.5,'nearest') ↵

fn =
      2
```

Now we get $f[m', n', k'] = 2$ using the nearest neighborhood interpolation method in the given 3D space.

Above execution brings an end to this chapter.

Mohammad Nuruzzaman

Exercises

1. An area over the domain $-0.9cm \le x \le 4.5cm$ and $-1.6cm \le y \le 5.6cm$ is chosen to apply the FD technique with $\Delta x = 0.9$ cm and $\Delta y = 0.8$ cm. What is the FD domain in integer coordinates? How many sample coordinates are in the domain? Do the same if each resolution is reduced to half.

2. Suppose a discrete two dimensional function $f[m,n]$ varies subject to $\Delta x = 0.5$ cm and $\Delta y = 0.6$ cm over the integer domain $-3 \le m \le 8$ and $-5 \le n \le 11$. What is the continuous domain? Do the same if the resolution is reduced to half.

3. Suppose $f(x,y) = 3xy^3 - x^3$ is to be computed based on FD method subject to $\Delta x = 0.3$ and $\Delta y = 0.5$ over $-0.5 \le x \le 0.7$ and $-1 \le y \le 1$. What are the grid point matrices? Which mathematical operation provides $f(x,y)$ samples employing the grid point matrices? Compute the samples employing the grid point matrices.

4. In question (3) determine f_{min}, f_{max}, and $f[m,n]$ if four levels are chosen as functional resolution. What are the FD reconstructed $\hat{f}(x,y)$ samples? Calculate the mean square error in the FD approximation. What will be the $\hat{f}(x,y)$ samples in conventional coordinate system?

$$\begin{pmatrix} 1.1 & 1.1 & 0.69 & 1 & 1 \\ 1 & 0.65 & 0.6 & 0.65 & 1 \\ 1 & 0.87 & 1 & 0.8 & 1 \\ 1 & 0.81 & 1.2 & 1 & 1 \\ 1.1 & 1.1 & 1 & 0.8 & 1 \\ 0.9 & 0.8 & 1.01 & 0.8 & 1 \\ 0.8 & 1 & 0.89 & 0.8 & 1 \\ 0.8 & 0.8 & 1 & 0.8 & 0.8 \end{pmatrix}$$

Figure E.2(a) Some $f(x,y)$ samples

5. In question (4) now the $f(x,y)$ samples are taken from the figure E.2(a). With $\Delta x = 0.4$ and $\Delta y = 0.6$, what is the $f(x,y)$ domain based on the convention of figure 2.1(a)?

6. Obtain the excel file Data5.xls through the email link of page ii and place the file in your working path of MATLAB. Make sure the data in file looks like figure E.2(b). In question (4) now $f(x,y)$ samples are taken from the

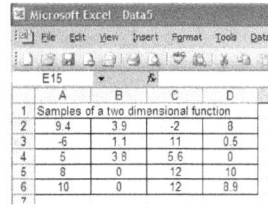

	A	B	C	D
1	Samples of a two dimensional function			
2	9.4	3.9	-2	8
3	-6	1.1	11	0.5
4	5	3.8	5.6	0
5	8	0	12	10
6	10	0	12	8.9

Figure E.2(b) Some $f(x,y)$ samples in an excel file

Figure E.2(c) Samples of figure E.2(b) with domain information - right side figure

	A	B	C	D	E	F
1	Samples of a two dimensional function				Horizontal	Vertical
2	9.4	3.9	-2	8	0	0
3	-6	1.1	11	0.5	1.5	1
4	5	3.8	5.6	0	3	2
5	8	0	12	10	4.5	3
6	10	0	12	8.9		4

figure E.2(b). With $\Delta x = 0.3$ and $\Delta y = 0.5$, what is the $f(x, y)$ domain based on the convention of figure 2.1(a)?

7. What if the domain information in question (6) is given as in figure E.2(c)? Obtain the excel file **Data6.xls** which holds the data of figure E.2(c).

8. Generate random samples for $f(x, y)$ each of uniform distribution within $-7V$ and $15V$ with $\Delta x = 0.2mm$ and $\Delta y = 0.3mm$ over $0 \le x \le 8mm$ and $0 \le y \le 6mm$. Perform question (4) quoted FD computing on the random samples of $f(x, y)$ based on 12 level functional resolution.

9. Contemplate about the grid point matrices and write MATLAB codes to compute the following double summations by using FD method: (a)

$$\sum_{q=-1}^{3} \sum_{p=-2}^{4} (2qp^2 - 5pq) \quad \text{(b)} \ \sum_{q=-1}^{3} \sum_{p=-2}^{4} (p^2 - 10) \quad \text{(c)} \ \sum_{q=0}^{4} \sum_{p=-5}^{5} \frac{\pi}{(p+1)(q-1)}, \ p \ne -1,$$

$q \ne 1$ (d) $\sum_{n=-2}^{4} \sum_{m=-3}^{2} f(m\Delta x, n\Delta y)$ over $-1 \le x \le 1$ and $0 \le y \le 1.2$ where

$f(x, y) = -3x^2 - 6yx + 12y$ (e) $\sum_{y} \sum_{x} f(n\Delta x, n\Delta y)$ over $0 \le x \le 1$ and $0 \le y \le 2$

where $f(x, y) = y^2 - 5yx$ with $\Delta x = 0.05$ and $\Delta y = 0.1$.

10. Applying 3D array basis compute $f(x, y, z) = \ln|0.1 + 10x - 2y^2 - 3z|$ over the domain $-3 \le x \le 2$, $-1 \le y \le 1$, and $-2 \le z \le 0$ subject to the resolutions $\Delta x = 1$, $\Delta y = 0.5$, and $\Delta z = 1$ in MATLAB.

11. In question (10) determine f_{min}, f_{max}, and $f[m, n, k]$ if five levels are chosen as functional resolution. What are the FD reconstructed $\hat{f}(x, y, z)$ samples? Calculate the mean square error in the FD approximation.

12. In question (11) determine m, n, and k directed derivatives on $f[m, n, k]$. Also determine the divided differences G_x, G_y, and G_z on $f(x, y, z)$, exercise theory defined forward gradient only.

13. In question (11) apply 3D bilinear interpolation on $f[m, n, k]$ when each resolution is reduced to half and determine the $f[m', n', k']$.

14. In question (13) now nearest neighborhood scheme is chosen.

Answers:

(1) $-1 \leq m \leq 5$, $-2 \leq n \leq 7$, 70, $-2 \leq m \leq 10$, $-4 \leq n \leq 14$, and 247 respectively

hint: section 2.1

(2) $-1.5cm \leq x \leq 4cm$ and $-3cm \leq y \leq 6.6cm$ and $-0.75cm \leq x \leq 2cm$ and $-1.5cm \leq y \leq 3.3cm$ respectively   hint: section 2.1

(3) $X = \begin{bmatrix} -0.5 & -0.2 & 0.1 & 0.4 & 0.7 \\ -0.5 & -0.2 & 0.1 & 0.4 & 0.7 \\ -0.5 & -0.2 & 0.1 & 0.4 & 0.7 \\ -0.5 & -0.2 & 0.1 & 0.4 & 0.7 \\ -0.5 & -0.2 & 0.1 & 0.4 & 0.7 \end{bmatrix}$ and $Y = \begin{bmatrix} -1 & -1 & -1 & -1 & -1 \\ -0.5 & -0.5 & -0.5 & -0.5 & -0.5 \\ 0 & 0 & 0 & 0 & 0 \\ 0.5 & 0.5 & 0.5 & 0.5 & 0.5 \\ 1 & 1 & 1 & 1 & 1 \end{bmatrix}$

$f(X,Y) = 3XY^3 - X^3$

$f(x,y)$ samples: $\begin{bmatrix} 1.6250 & 0.6080 & -0.3010 & -1.2640 & -2.4430 \\ 0.3125 & 0.0830 & -0.0385 & -0.2140 & -0.6055 \\ 0.1250 & 0.0080 & -0.0010 & -0.0640 & -0.3430 \\ -0.0625 & -0.0670 & 0.0365 & 0.0860 & -0.0805 \\ -1.3750 & -0.5920 & 0.2990 & 1.1360 & 1.7570 \end{bmatrix}$

hint: section 2.2

(4) $f_{min} = -2.443$, $f_{max} = 1.757$, $\Delta f = 1.4$, and $f[m,n] = \begin{bmatrix} 3 & 2 & 2 & 1 & 0 \\ 2 & 2 & 2 & 2 & 1 \\ 2 & 2 & 2 & 2 & 2 \\ 2 & 2 & 2 & 2 & 2 \\ 1 & 1 & 2 & 3 & 3 \end{bmatrix}$

$\hat{f}(x,y) = \begin{bmatrix} 1.7570 & 0.3570 & 0.3570 & -1.0430 & -2.4430 \\ 0.3570 & 0.3570 & 0.3570 & 0.3570 & -1.0430 \\ 0.3570 & 0.3570 & 0.3570 & 0.3570 & 0.3570 \\ 0.3570 & 0.3570 & 0.3570 & 0.3570 & 0.3570 \\ -1.0430 & -1.0430 & 0.3570 & 1.7570 & 1.7570 \end{bmatrix}$

mse=0.1484

$\hat{f}(x,y) = \begin{bmatrix} -1.0430 & -1.0430 & 0.3570 & 1.7570 & 1.7570 \\ 0.3570 & 0.3570 & 0.3570 & 0.3570 & 0.3570 \\ 0.3570 & 0.3570 & 0.3570 & 0.3570 & 0.3570 \\ 0.3570 & 0.3570 & 0.3570 & 0.3570 & -1.0430 \\ 1.7570 & 0.3570 & 0.3570 & -1.0430 & -2.4430 \end{bmatrix}$

hint: section 2.2

(5) $f_{min} = 0.6$, $f_{max} = 1.2$, $\Delta f = 0.2$, and $f[m,n] = \begin{bmatrix} 3 & 3 & 0 & 2 & 2 \\ 2 & 0 & 0 & 0 & 2 \\ 2 & 1 & 2 & 1 & 2 \\ 2 & 1 & 3 & 2 & 2 \\ 3 & 3 & 2 & 1 & 2 \\ 2 & 1 & 2 & 1 & 2 \\ 1 & 2 & 1 & 1 & 2 \\ 1 & 1 & 2 & 1 & 1 \end{bmatrix}$

$$\hat{f}(x,y) \;=\; \begin{bmatrix} 1.2 & 1.2 & 0.6 & 1 & 1 \\ 1 & 0.6 & 0.6 & 0.6 & 1 \\ 1 & 0.8 & 1 & 0.8 & 1 \\ 1 & 0.8 & 1.2 & 1 & 1 \\ 1.2 & 1.2 & 1 & 0.8 & 1 \\ 1 & 0.8 & 1 & 0.8 & 1 \\ 0.8 & 1 & 0.8 & 0.8 & 1 \\ 0.8 & 0.8 & 1 & 0.8 & 0.8 \end{bmatrix}$$

mse=0.0019, $0 \le x \le 1.6$ and $0 \le y \le 4.2$ assuming 0 start of x and y

(6) f_{min} =−6, f_{max} =12, Δf =6, and $f[m,n] = \begin{bmatrix} 3 & 2 & 1 & 2 \\ 0 & 1 & 3 & 1 \\ 2 & 2 & 2 & 1 \\ 2 & 1 & 3 & 3 \\ 3 & 1 & 3 & 2 \end{bmatrix}$

$$\hat{f}(x,y) \;=\; \begin{bmatrix} 12 & 6 & 0 & 6 \\ -6 & 0 & 12 & 0 \\ 6 & 6 & 6 & 0 \\ 6 & 0 & 12 & 12 \\ 12 & 0 & 12 & 6 \end{bmatrix}$$

mse=2.402, $0 \le x \le 0.9$ and $0 \le y \le 2$ assuming 0 start of x and y

(7) f_{min} , f_{max} , Δf , $f[m,n]$, and $\hat{f}(x,y)$ are the same as those in question (6) just domain interval is different which is $0 \le x \le 4.5$ and $0 \le y \le 4$ with $\Delta x = 1.5$ and $\Delta y = 1$. Also get all data by executing first D=xlsread('Data6.xls'); and then f=D(:,1:4); for the $f(x,y)$ samples where D is a user-chosen variable.

(8) For randomness we are not going to get unique values. Codes: M=8/0.2+1; N=6/0.3+1; f=22*rand(N,M)-7; df=(15-(-7))/(12-1); fmn=round((f-(-7))/df); f_hat=fmn*df+min(f(:)); mse(f-f_hat) where the symbols have their usual meanings and assuming level variation between given minimum and maximum.

(9) (a)

$$P \;=\; \begin{bmatrix} -2 & -1 & 0 & 1 & 2 & 3 & 4 \\ -2 & -1 & 0 & 1 & 2 & 3 & 4 \\ -2 & -1 & 0 & 1 & 2 & 3 & 4 \\ -2 & -1 & 0 & 1 & 2 & 3 & 4 \\ -2 & -1 & 0 & 1 & 2 & 3 & 4 \end{bmatrix}$$

$$Q \;=\; \begin{bmatrix} -1 & -1 & -1 & -1 & -1 & -1 & -1 \\ 0 & 0 & 0 & 0 & 0 & 0 & 0 \\ 1 & 1 & 1 & 1 & 1 & 1 & 1 \\ 2 & 2 & 2 & 2 & 2 & 2 & 2 \\ 3 & 3 & 3 & 3 & 3 & 3 & 3 \end{bmatrix}$$

175
(b) −175 (c) 0.9599 (d) 243.6 (e) −499.8

Hint: sections 2.3 and 2.4 and appendix B.9

(10)

$$f(x,y,-2) = \begin{bmatrix} 3.2542 & 2.7663 & 1.7750 & 1.4110 & 2.6462 & 3.1822 \\ 3.1946 & 2.6672 & 1.4816 & 1.7228 & 2.7473 & 3.2426 \\ 3.1739 & 2.6319 & 1.3610 & 1.8083 & 2.7788 & 3.2619 \\ 3.1946 & 2.6672 & 1.4816 & 1.7228 & 2.7473 & 3.2426 \\ 3.2542 & 2.7663 & 1.7750 & 1.4110 & 2.6462 & 3.1822 \end{bmatrix}$$

$$f(x,y,-1) = \begin{bmatrix} 3.3638 & 2.9392 & 2.1861 & 0.0953 & 2.4069 & 3.0493 \\ 3.3105 & 2.8565 & 2.0015 & 0.9555 & 2.5337 & 3.1179 \\ 3.2921 & 2.8273 & 1.9315 & 1.1314 & 2.5726 & 3.1398 \\ 3.3105 & 2.8565 & 2.0015 & 0.9555 & 2.5337 & 3.1179 \\ 3.3638 & 2.9392 & 2.1861 & 0.0953 & 2.4069 & 3.0493 \end{bmatrix}$$

$$f(x,y,0) = \begin{bmatrix} 3.4626 & 3.0865 & 2.4765 & 0.6419 & 2.0919 & 2.8959 \\ 3.4144 & 3.0155 & 2.3418 & -0.9163 & 2.2618 & 2.9755 \\ 3.3979 & 2.9907 & 2.2925 & -2.3026 & 2.3125 & 3.0007 \\ 3.4144 & 3.0155 & 2.3418 & -0.9163 & 2.2618 & 2.9755 \\ 3.4626 & 3.0865 & 2.4765 & 0.6419 & 2.0919 & 2.8959 \end{bmatrix}$$

Hint: section 2.6

(11) $f_{\min} = -2.3026$ and $f_{\max} = 3.4626$

$$f[m,n,-2] = \begin{bmatrix} 4 & 4 & 3 & 3 & 3 & 4 \\ 4 & 3 & 3 & 3 & 4 & 4 \\ 4 & 3 & 3 & 3 & 4 & 4 \\ 4 & 3 & 3 & 3 & 4 & 4 \\ 4 & 4 & 3 & 3 & 3 & 4 \end{bmatrix}$$

$$f[m,n,-1] = \begin{bmatrix} 4 & 4 & 3 & 2 & 3 & 4 \\ 4 & 4 & 3 & 2 & 3 & 4 \\ 4 & 4 & 3 & 2 & 3 & 4 \\ 4 & 4 & 3 & 2 & 3 & 4 \\ 4 & 4 & 3 & 2 & 3 & 4 \end{bmatrix}$$

$$f[m,n,0] = \begin{bmatrix} 4 & 4 & 3 & 2 & 3 & 4 \\ 4 & 4 & 3 & 1 & 3 & 4 \\ 4 & 4 & 3 & 0 & 3 & 4 \\ 4 & 4 & 3 & 1 & 3 & 4 \\ 4 & 4 & 3 & 2 & 3 & 4 \end{bmatrix}$$

$$\hat{f}(x,y,-2) = \begin{bmatrix} 3.4626 & 3.4626 & 2.0213 & 2.0213 & 2.0213 & 3.4626 \\ 3.4626 & 2.0213 & 2.0213 & 2.0213 & 3.4626 & 3.4626 \\ 3.4626 & 2.0213 & 2.0213 & 2.0213 & 3.4626 & 3.4626 \\ 3.4626 & 2.0213 & 2.0213 & 2.0213 & 3.4626 & 3.4626 \\ 3.4626 & 3.4626 & 2.0213 & 2.0213 & 2.0213 & 3.4626 \end{bmatrix}$$

$$\hat{f}(x,y,-1) = \begin{bmatrix} 3.4626 & 3.4626 & 2.0213 & 0.5800 & 2.0213 & 3.4626 \\ 3.4626 & 3.4626 & 2.0213 & 0.5800 & 2.0213 & 3.4626 \\ 3.4626 & 3.4626 & 2.0213 & 0.5800 & 2.0213 & 3.4626 \\ 3.4626 & 3.4626 & 2.0213 & 0.5800 & 2.0213 & 3.4626 \\ 3.4626 & 3.4626 & 2.0213 & 0.5800 & 2.0213 & 3.4626 \end{bmatrix}$$

$$\hat{f}(x,y,0) = \begin{bmatrix} 3.4626 & 3.4626 & 2.0213 & 0.5800 & 2.0213 & 3.4626 \\ 3.4626 & 3.4626 & 2.0213 & -0.8613 & 2.0213 & 3.4626 \\ 3.4626 & 3.4626 & 2.0213 & -2.3026 & 2.0213 & 3.4626 \\ 3.4626 & 3.4626 & 2.0213 & -0.8613 & 2.0213 & 3.4626 \\ 3.4626 & 3.4626 & 2.0213 & 0.5800 & 2.0213 & 3.4626 \end{bmatrix}$$

Mean square error: 0.1671   Hint: sections 2.5 and 2.6

(12)

$$G_m[m,n,-2] = \begin{bmatrix} 0 & -1 & 0 & 0 & 1 \\ -1 & 0 & 0 & 1 & 0 \\ -1 & 0 & 0 & 1 & 0 \\ -1 & 0 & 0 & 1 & 0 \\ 0 & -1 & 0 & 0 & 1 \end{bmatrix}$$

$$G_m[m,n,-1] = \begin{bmatrix} 0 & -1 & -1 & 1 & 1 \\ 0 & -1 & -1 & 1 & 1 \\ 0 & -1 & -1 & 1 & 1 \\ 0 & -1 & -1 & 1 & 1 \\ 0 & -1 & -1 & 1 & 1 \end{bmatrix}$$

$$G_m[m,n,0] = \begin{bmatrix} 0 & -1 & -1 & 1 & 1 \\ 0 & -1 & -2 & 2 & 1 \\ 0 & -1 & -3 & 3 & 1 \\ 0 & -1 & -2 & 2 & 1 \\ 0 & -1 & -1 & 1 & 1 \end{bmatrix}$$

$$G_n[m,n,-2] = \begin{bmatrix} 0 & -1 & 0 & 0 & 1 & 0 \\ 0 & 0 & 0 & 0 & 0 & 0 \\ 0 & 0 & 0 & 0 & 0 & 0 \\ 0 & 1 & 0 & 0 & -1 & 0 \end{bmatrix}$$

$$G_n[m,n,-1] = \begin{bmatrix} 0 & 0 & 0 & 0 & 0 & 0 \\ 0 & 0 & 0 & 0 & 0 & 0 \\ 0 & 0 & 0 & 0 & 0 & 0 \\ 0 & 0 & 0 & 0 & 0 & 0 \end{bmatrix}$$

$$G_n[m,n,0] = \begin{bmatrix} 0 & 0 & 0 & -1 & 0 & 0 \\ 0 & 0 & 0 & -1 & 0 & 0 \\ 0 & 0 & 0 & 1 & 0 & 0 \\ 0 & 0 & 0 & 1 & 0 & 0 \end{bmatrix}$$

$$G_k[m,n,-2] = \begin{bmatrix} 0 & 0 & 0 & -1 & 0 & 0 \\ 0 & 1 & 0 & -1 & -1 & 0 \\ 0 & 1 & 0 & -1 & -1 & 0 \\ 0 & 1 & 0 & -1 & -1 & 0 \\ 0 & 0 & 0 & -1 & 0 & 0 \end{bmatrix}$$

$$G_k[m,n,-1] = \begin{bmatrix} 0 & 0 & 0 & 0 & 0 & 0 \\ 0 & 0 & 0 & -1 & 0 & 0 \\ 0 & 0 & 0 & -2 & 0 & 0 \\ 0 & 0 & 0 & -1 & 0 & 0 \\ 0 & 0 & 0 & 0 & 0 & 0 \end{bmatrix}$$

$$G_x(x,y,-2) = \begin{bmatrix} -0.4879 & -0.9914 & -0.3640 & 1.2352 & 0.5360 \\ -0.5274 & -1.1856 & 0.2412 & 1.0245 & 0.4953 \\ -0.5420 & -1.2709 & 0.4473 & 0.9705 & 0.4831 \\ -0.5274 & -1.1856 & 0.2412 & 1.0245 & 0.4953 \\ -0.4879 & -0.9914 & -0.3640 & 1.2352 & 0.5360 \end{bmatrix}$$

$$G_x(x,y,-1) = \begin{bmatrix} -0.4247 & -0.7531 & -2.0907 & 2.3116 & 0.6423 \\ -0.4541 & -0.8550 & -1.0460 & 1.5782 & 0.5843 \\ -0.4648 & -0.8958 & -0.8001 & 1.4412 & 0.5672 \\ -0.4541 & -0.8550 & -1.0460 & 1.5782 & 0.5843 \\ -0.4247 & -0.7531 & -2.0907 & 2.3116 & 0.6423 \end{bmatrix}$$

$$G_x(x,y,0) = \begin{bmatrix} -0.3761 & -0.6099 & -1.8347 & 1.4500 & 0.8040 \\ -0.3989 & -0.6737 & -3.2581 & 3.1781 & 0.7138 \\ -0.4071 & -0.6982 & -4.5951 & 4.6151 & 0.6882 \\ -0.3989 & -0.6737 & -3.2581 & 3.1781 & 0.7138 \\ -0.3761 & -0.6099 & -1.8347 & 1.4500 & 0.8040 \end{bmatrix}$$

$$G_y(x,y,-2) = \begin{bmatrix} -0.1193 & -0.1982 & -0.5867 & 0.6236 & 0.2022 & 0.1208 \\ -0.0414 & -0.0707 & -0.2413 & 0.1710 & 0.0631 & 0.0387 \\ 0.0414 & 0.0707 & 0.2413 & -0.1710 & -0.0631 & -0.0387 \\ 0.1193 & 0.1982 & 0.5867 & -0.6236 & -0.2022 & -0.1208 \end{bmatrix}$$

$$G_y(x,y,-1) = \begin{bmatrix} -0.1066 & -0.1654 & -0.3691 & 1.7204 & 0.2535 & 0.1374 \\ -0.0368 & -0.0583 & -0.1399 & 0.3518 & 0.0778 & 0.0438 \\ 0.0368 & 0.0583 & 0.1399 & -0.3518 & -0.0778 & -0.0438 \\ 0.1066 & 0.1654 & 0.3691 & -1.7204 & -0.2535 & -0.1374 \end{bmatrix}$$

$$G_y(x,y,0) = \begin{bmatrix} -0.0963 & -0.1419 & -0.2695 & -3.1163 & 0.3398 & 0.1592 \\ -0.0332 & -0.0496 & -0.0985 & -2.7726 & 0.1015 & 0.0504 \\ 0.0332 & 0.0496 & 0.0985 & 2.7726 & -0.1015 & -0.0504 \\ 0.0963 & 0.1419 & 0.2695 & 3.1163 & -0.3398 & -0.1592 \end{bmatrix}$$

$$G_z(x,y,-2) = \begin{bmatrix} 0.1096 & 0.1728 & 0.4111 & -1.3157 & -0.2392 & -0.1329 \\ 0.1160 & 0.1892 & 0.5199 & -0.7673 & -0.2136 & -0.1246 \\ 0.1182 & 0.1954 & 0.5705 & -0.6769 & -0.2062 & -0.1221 \\ 0.1160 & 0.1892 & 0.5199 & -0.7673 & -0.2136 & -0.1246 \\ 0.1096 & 0.1728 & 0.4111 & -1.3157 & -0.2392 & -0.1329 \end{bmatrix}$$

$$G_z(x,y,-1) = \begin{bmatrix} 0.0988 & 0.1473 & 0.2905 & 0.5465 & -0.3151 & -0.1534 \\ 0.1039 & 0.1591 & 0.3403 & -1.8718 & -0.2719 & -0.1424 \\ 0.1057 & 0.1634 & 0.3610 & -3.4340 & -0.2601 & -0.1391 \\ 0.1039 & 0.1591 & 0.3403 & -1.8718 & -0.2719 & -0.1424 \\ 0.0988 & 0.1473 & 0.2905 & 0.5465 & -0.3151 & -0.1534 \end{bmatrix}$$

Hint: sections 2.7 and 2.8

(13) $f[m',n',-2] =$

4.0000	4.0000	4.0000	3.5000	3.0000	3.0000	3.0000	3.0000	3.0000	3.5000	4.0000
4.0000	3.7500	3.5000	3.2500	3.0000	3.0000	3.0000	3.2500	3.5000	3.7500	4.0000
4.0000	3.5000	3.0000	3.0000	3.0000	3.0000	3.0000	3.5000	4.0000	4.0000	4.0000
4.0000	3.5000	3.0000	3.0000	3.0000	3.0000	3.0000	3.5000	4.0000	4.0000	4.0000
4.0000	3.5000	3.0000	3.0000	3.0000	3.0000	3.0000	3.5000	4.0000	4.0000	4.0000
4.0000	3.5000	3.0000	3.0000	3.0000	3.0000	3.0000	3.5000	4.0000	4.0000	4.0000
4.0000	3.5000	3.0000	3.0000	3.0000	3.0000	3.0000	3.5000	4.0000	4.0000	4.0000
4.0000	3.7500	3.5000	3.2500	3.0000	3.0000	3.0000	3.2500	3.5000	3.7500	4.0000
4.0000	4.0000	4.0000	3.5000	3.0000	3.0000	3.0000	3.0000	3.0000	3.5000	4.0000

$f[m',n',-1.5] =$

4.0000	4.0000	4.0000	3.5000	3.0000	2.7500	2.5000	2.7500	3.0000	3.5000	4.0000
4.0000	3.8750	3.7500	3.3750	3.0000	2.7500	2.5000	2.8750	3.2500	3.6250	4.0000
4.0000	3.7500	3.5000	3.2500	3.0000	2.7500	2.5000	3.0000	3.5000	3.7500	4.0000
4.0000	3.7500	3.5000	3.2500	3.0000	2.7500	2.5000	3.0000	3.5000	3.7500	4.0000
4.0000	3.7500	3.5000	3.2500	3.0000	2.7500	2.5000	3.0000	3.5000	3.7500	4.0000
4.0000	3.7500	3.5000	3.2500	3.0000	2.7500	2.5000	3.0000	3.5000	3.7500	4.0000
4.0000	3.7500	3.5000	3.2500	3.0000	2.7500	2.5000	3.0000	3.5000	3.7500	4.0000
4.0000	3.8750	3.7500	3.3750	3.0000	2.7500	2.5000	2.8750	3.2500	3.6250	4.0000
4.0000	4.0000	4.0000	3.5000	3.0000	2.7500	2.5000	2.7500	3.0000	3.5000	4.0000

$f[m',n',-1] =$

$$
\begin{bmatrix}
4.0000 & 4.0000 & 4.0000 & 3.5000 & 3.0000 & 2.5000 & 2.0000 & 2.5000 & 3.0000 & 3.5000 & 4.0000 \\
4.0000 & 4.0000 & 4.0000 & 3.5000 & 3.0000 & 2.3750 & 1.7500 & 2.3750 & 3.0000 & 3.5000 & 4.0000 \\
4.0000 & 4.0000 & 4.0000 & 3.5000 & 3.0000 & 2.2500 & 1.5000 & 2.2500 & 3.0000 & 3.5000 & 4.0000 \\
4.0000 & 4.0000 & 4.0000 & 3.5000 & 3.0000 & 2.1250 & 1.2500 & 2.1250 & 3.0000 & 3.5000 & 4.0000 \\
4.0000 & 4.0000 & 4.0000 & 3.5000 & 3.0000 & 2.0000 & 1.0000 & 2.0000 & 3.0000 & 3.5000 & 4.0000 \\
4.0000 & 4.0000 & 4.0000 & 3.5000 & 3.0000 & 2.1250 & 1.2500 & 2.1250 & 3.0000 & 3.5000 & 4.0000 \\
4.0000 & 4.0000 & 4.0000 & 3.5000 & 3.0000 & 2.2500 & 1.5000 & 2.2500 & 3.0000 & 3.5000 & 4.0000 \\
4.0000 & 4.0000 & 4.0000 & 3.5000 & 3.0000 & 2.3750 & 1.7500 & 2.3750 & 3.0000 & 3.5000 & 4.0000 \\
4.0000 & 4.0000 & 4.0000 & 3.5000 & 3.0000 & 2.5000 & 2.0000 & 2.5000 & 3.0000 & 3.5000 & 4.0000 \\
\end{bmatrix}
$$

$f[m',n',-0.5] =$

$$
\begin{bmatrix}
4.0000 & 4.0000 & 4.0000 & 3.5000 & 3.0000 & 2.5000 & 2.0000 & 2.5000 & 3.0000 & 3.5000 & 4.0000 \\
4.0000 & 4.0000 & 4.0000 & 3.5000 & 3.0000 & 2.3750 & 1.7500 & 2.3750 & 3.0000 & 3.5000 & 4.0000 \\
4.0000 & 4.0000 & 4.0000 & 3.5000 & 3.0000 & 2.2500 & 1.5000 & 2.2500 & 3.0000 & 3.5000 & 4.0000 \\
4.0000 & 4.0000 & 4.0000 & 3.5000 & 3.0000 & 2.1250 & 1.2500 & 2.1250 & 3.0000 & 3.5000 & 4.0000 \\
4.0000 & 4.0000 & 4.0000 & 3.5000 & 3.0000 & 2.0000 & 1.0000 & 2.0000 & 3.0000 & 3.5000 & 4.0000 \\
4.0000 & 4.0000 & 4.0000 & 3.5000 & 3.0000 & 2.1250 & 1.2500 & 2.1250 & 3.0000 & 3.5000 & 4.0000 \\
4.0000 & 4.0000 & 4.0000 & 3.5000 & 3.0000 & 2.2500 & 1.5000 & 2.2500 & 3.0000 & 3.5000 & 4.0000 \\
4.0000 & 4.0000 & 4.0000 & 3.5000 & 3.0000 & 2.3750 & 1.7500 & 2.3750 & 3.0000 & 3.5000 & 4.0000 \\
4.0000 & 4.0000 & 4.0000 & 3.5000 & 3.0000 & 2.5000 & 2.0000 & 2.5000 & 3.0000 & 3.5000 & 4.0000 \\
\end{bmatrix}
$$

$f[m',n',0] =$

$$
\begin{bmatrix}
4.0000 & 4.0000 & 4.0000 & 3.5000 & 3.0000 & 2.5000 & 2.0000 & 2.5000 & 3.0000 & 3.5000 & 4.0000 \\
4.0000 & 4.0000 & 4.0000 & 3.5000 & 3.0000 & 2.2500 & 1.5000 & 2.2500 & 3.0000 & 3.5000 & 4.0000 \\
4.0000 & 4.0000 & 4.0000 & 3.5000 & 3.0000 & 2.0000 & 1.0000 & 2.0000 & 3.0000 & 3.5000 & 4.0000 \\
4.0000 & 4.0000 & 4.0000 & 3.5000 & 3.0000 & 1.7500 & 0.5000 & 1.7500 & 3.0000 & 3.5000 & 4.0000 \\
4.0000 & 4.0000 & 4.0000 & 3.5000 & 3.0000 & 1.5000 & 0 & 1.5000 & 3.0000 & 3.5000 & 4.0000 \\
4.0000 & 4.0000 & 4.0000 & 3.5000 & 3.0000 & 1.7500 & 0.5000 & 1.7500 & 3.0000 & 3.5000 & 4.0000 \\
4.0000 & 4.0000 & 4.0000 & 3.5000 & 3.0000 & 2.0000 & 1.0000 & 2.0000 & 3.0000 & 3.5000 & 4.0000 \\
4.0000 & 4.0000 & 4.0000 & 3.5000 & 3.0000 & 2.2500 & 1.5000 & 2.2500 & 3.0000 & 3.5000 & 4.0000 \\
4.0000 & 4.0000 & 4.0000 & 3.5000 & 3.0000 & 2.5000 & 2.0000 & 2.5000 & 3.0000 & 3.5000 & 4.0000 \\
\end{bmatrix}
$$

Hint: sections 2.9 and 2.10

(14) Nearest neighborhood interpolation:

$$
f[m',n',-2] =
\begin{bmatrix}
4 & 4 & 4 & 3 & 3 & 3 & 3 & 3 & 3 & 4 & 4 \\
4 & 3 & 3 & 3 & 3 & 3 & 3 & 4 & 4 & 4 & 4 \\
4 & 3 & 3 & 3 & 3 & 3 & 3 & 4 & 4 & 4 & 4 \\
4 & 3 & 3 & 3 & 3 & 3 & 3 & 4 & 4 & 4 & 4 \\
4 & 3 & 3 & 3 & 3 & 3 & 3 & 4 & 4 & 4 & 4 \\
4 & 3 & 3 & 3 & 3 & 3 & 3 & 4 & 4 & 4 & 4 \\
4 & 3 & 3 & 3 & 3 & 3 & 3 & 4 & 4 & 4 & 4 \\
4 & 4 & 4 & 3 & 3 & 3 & 3 & 3 & 3 & 4 & 4 \\
4 & 4 & 4 & 3 & 3 & 3 & 3 & 3 & 3 & 4 & 4 \\
\end{bmatrix}
$$

$$
f[m',n',-1.5] =
\begin{bmatrix}
4 & 4 & 4 & 3 & 3 & 2 & 2 & 3 & 3 & 4 & 4 \\
4 & 4 & 4 & 3 & 3 & 2 & 2 & 3 & 3 & 4 & 4 \\
4 & 4 & 4 & 3 & 3 & 2 & 2 & 3 & 3 & 4 & 4 \\
4 & 4 & 4 & 3 & 3 & 2 & 2 & 3 & 3 & 4 & 4 \\
4 & 4 & 4 & 3 & 3 & 2 & 2 & 3 & 3 & 4 & 4 \\
4 & 4 & 4 & 3 & 3 & 2 & 2 & 3 & 3 & 4 & 4 \\
4 & 4 & 4 & 3 & 3 & 2 & 2 & 3 & 3 & 4 & 4 \\
4 & 4 & 4 & 3 & 3 & 2 & 2 & 3 & 3 & 4 & 4 \\
4 & 4 & 4 & 3 & 3 & 2 & 2 & 3 & 3 & 4 & 4 \\
\end{bmatrix}
$$

$$
f[m',n',-1] = \begin{bmatrix}
4 & 4 & 4 & 3 & 3 & 2 & 2 & 3 & 3 & 4 & 4 \\
4 & 4 & 4 & 3 & 3 & 2 & 2 & 3 & 3 & 4 & 4 \\
4 & 4 & 4 & 3 & 3 & 2 & 2 & 3 & 3 & 4 & 4 \\
4 & 4 & 4 & 3 & 3 & 2 & 2 & 3 & 3 & 4 & 4 \\
4 & 4 & 4 & 3 & 3 & 2 & 2 & 3 & 3 & 4 & 4 \\
4 & 4 & 4 & 3 & 3 & 2 & 2 & 3 & 3 & 4 & 4 \\
4 & 4 & 4 & 3 & 3 & 2 & 2 & 3 & 3 & 4 & 4 \\
4 & 4 & 4 & 3 & 3 & 2 & 2 & 3 & 3 & 4 & 4 \\
4 & 4 & 4 & 3 & 3 & 2 & 2 & 3 & 3 & 4 & 4
\end{bmatrix}
$$

$$
f[m',n',-0.5] = \begin{bmatrix}
4 & 4 & 4 & 3 & 3 & 2 & 2 & 3 & 3 & 4 & 4 \\
4 & 4 & 4 & 3 & 3 & 1 & 1 & 3 & 3 & 4 & 4 \\
4 & 4 & 4 & 3 & 3 & 1 & 1 & 3 & 3 & 4 & 4 \\
4 & 4 & 4 & 3 & 3 & 0 & 0 & 3 & 3 & 4 & 4 \\
4 & 4 & 4 & 3 & 3 & 0 & 0 & 3 & 3 & 4 & 4 \\
4 & 4 & 4 & 3 & 3 & 1 & 1 & 3 & 3 & 4 & 4 \\
4 & 4 & 4 & 3 & 3 & 1 & 1 & 3 & 3 & 4 & 4 \\
4 & 4 & 4 & 3 & 3 & 2 & 2 & 3 & 3 & 4 & 4 \\
4 & 4 & 4 & 3 & 3 & 2 & 2 & 3 & 3 & 4 & 4
\end{bmatrix}
$$

$$
f[m',n',0] = \begin{bmatrix}
4 & 4 & 4 & 3 & 3 & 2 & 2 & 3 & 3 & 4 & 4 \\
4 & 4 & 4 & 3 & 3 & 1 & 1 & 3 & 3 & 4 & 4 \\
4 & 4 & 4 & 3 & 3 & 1 & 1 & 3 & 3 & 4 & 4 \\
4 & 4 & 4 & 3 & 3 & 0 & 0 & 3 & 3 & 4 & 4 \\
4 & 4 & 4 & 3 & 3 & 0 & 0 & 3 & 3 & 4 & 4 \\
4 & 4 & 4 & 3 & 3 & 1 & 1 & 3 & 3 & 4 & 4 \\
4 & 4 & 4 & 3 & 3 & 1 & 1 & 3 & 3 & 4 & 4 \\
4 & 4 & 4 & 3 & 3 & 2 & 2 & 3 & 3 & 4 & 4 \\
4 & 4 & 4 & 3 & 3 & 2 & 2 & 3 & 3 & 4 & 4
\end{bmatrix}
$$

Mohammad Nuruzzaman

Chapter 3

$$\nabla^2 f = 0$$

Solving 3D Laplace Equation by FD

This chapter largely focuses simulation on the solution of Laplace equation in three dimensional finite difference (3D FD) context. Only rectangular coordinate is chosen but intermediatory expressions can easily lead to the solutions in other coordinates. Slight theoretical background is provided to define three dimensional molecular equation in Cartesian system so that solution finding is facilely exercised although this is not a text book. Illustrated below outlines the conceptual sequence addressed in this chapter:

♦♦ 3D FD molecular equation addressing the 2nd order basics
♦♦ Author written function file for solving 3D Laplace equation
♦♦ Elaboration on boundary conditions of Laplace equation in 3D
♦♦ Solutions of Laplace equation attributed to variety of boundary conditions ranging linear to planar

3.1 Basics of Laplacian in 3D context

The Laplacian is simply given by $\nabla^2 f$ but it has many implications. Sections 2.1-2.4 are the prerequisite for this section. The operator ∇ is coordinate dependent and we focus only on rectangular system i.e. $\nabla = \overline{a}_x \frac{\partial}{\partial x} + \overline{a}_y \frac{\partial}{\partial y} + \overline{a}_z \frac{\partial}{\partial z}$ in this section where $\overline{a}_x$ is the unit vector in the x

direction, so are the other two in y and z directions respectively. Laplace equation i.e. $\nabla^2 f = 0$ provides distribution of scalar f if there is no source element inside the 3D domain. The expansion of $\nabla^2 f = 0$ becomes $\dfrac{\partial^2 f}{\partial x^2} + \dfrac{\partial^2 f}{\partial y^2} + \dfrac{\partial^2 f}{\partial z^2} = 0$. Section 2.4 quoted notations and approximation of second order partial derivatives provide the following:

$$\frac{\partial^2 f}{\partial x^2} \approx \frac{f[m+1,n,k] + f[m-1,n,k] - 2f[m,n,k]}{(\Delta x)^2},$$

$$\frac{\partial^2 f}{\partial y^2} \approx \frac{f[m,n+1,k] + f[m,n-1,k] - 2f[m,n,k]}{(\Delta y)^2}, \text{ and}$$

$$\frac{\partial^2 f}{\partial z^2} \approx \frac{f[m,n,k+1] + f[m,n,k-1] - 2f[m,n,k]}{(\Delta z)^2}.$$

Substituting the above in $\nabla^2 f = 0$ and solving for $f[m,n,k]$ get us

$$f[m,n,k] = \frac{\dfrac{f[m+1,n,k] + f[m-1,n,k]}{(\Delta x)^2} + \dfrac{f[m,n+1,k] + f[m,n-1,k]}{(\Delta y)^2} + \dfrac{f[m,n,k+1] + f[m,n,k-1]}{(\Delta z)^2}}{2\left\{\dfrac{1}{(\Delta x)^2} + \dfrac{1}{(\Delta y)^2} + \dfrac{1}{(\Delta z)^2}\right\}}.$$

Above equation is called the 3D finite difference counterpart of Laplace equation in rectangular system which we implement by iterative programming means in next section.

Along the x direction the discrete functional elements $f[m+1,n,k]$, $f[m-1,n,k]$, and $f[m,n,k]$ correspond to right (R), left (L), and center (C) grid points respectively i.e.

$$\begin{array}{ccc} L & C & R \\ \bullet \leftarrow \Delta x \rightarrow & \bullet \leftarrow \Delta x \rightarrow & \bullet \end{array}$$

In above every bold dot indicates a grid point in discrete domain. Along the y direction the discrete functional elements $f[m,n+1,k]$, $f[m,n-1,k]$, and $f[m,n,k]$ correspond to down (D), up (U), and center (C) grid points respectively (as far as coordinate system in figure 2.3(a) is concerned) i.e.

$$\begin{array}{cl} \bullet & U \\ \uparrow & \\ \Delta y & \\ \downarrow & \\ \bullet & C \\ \uparrow & \\ \Delta y & \\ \downarrow & \\ \bullet & D \end{array}$$

Along the z direction the discrete functional elements $f[m,n,k+1]$, $f[m,n,k-1]$, and $f[m,n,k]$ correspond to back (B), front (F), and center (C) grid points respectively i.e.

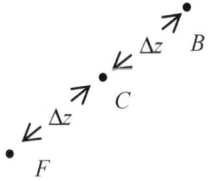

If we place all three above in three dimensional perspective, we obtain the finite difference molecule as follows:

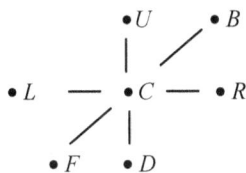

3.2 Solution of 3D Laplace equation using FD technique

Last section mentioned finite difference solution for $f[m,n,k]$ needs writing equations for every node and solving them subject to some boundary conditions. The problem is there will be so many equations for fine resolution (i.e. smaller Δx, Δy, or Δz) which becomes very clumsy as far as 3D context is involved.

Another approach is inject boundary condition to boundary grid points, compute $f[m,n,k]$ for every inside grid point, and repeat the computing until mean square error is negligible.

Unfortunately there is no embedded function in MATLAB which solves Laplace equation in 3D. Author written function lap3d can be an alternative. In this section we discuss how to utilize the function in order to obtain the solution of Laplace equation using three dimensional finite difference techniques.

In order to exercise the function the user has to decide three dimensional domain bounded by $x_1 \le x \le x_2$, $y_1 \le y \le y_2$, and $z_1 \le z \le z_2$. The first input argument of the lap3d is the domain bounds as a six element row matrix i.e. $[x_1, x_2, y_1, y_2, z_1, z_2]$.

The second, third, and fourth input arguments of the lap3d are the resolutions Δx, Δy, and Δz respectively.

The fifth input argument of lap3d is the boundary condition but as a string and we defined the solution variable by f which is a three dimensional array. Any boundary condition must oblige MATLAB indexing subject to the resolution Δx, Δy, or Δz. Suppose we wish to enter the boundary condition $f(0,0,0)=2$. In MATLAB array indexing starts from 1 so we write 'f(1,1,1)=2;' for the condition assuming that each bound starts from 0.

The given bound also matters. Say the 3D domain is defined by $1 \le x \le 3$, $0 \le y \le 2$, and $-1 \le z \le 1$ where $\Delta x=1$, $\Delta y=0.5$, and $\Delta z=1$ and the condition is $f(2,1,0)=2$. For x, 2 is the 2nd index in f (section 2.3 for more discussion). Again for y, 1 is the 3rd index in f so is 2 for the z thus we write the condition as 'f(2,3,2)=2;'.

The condition $f(2,1,z)=2$ indicates all elements of z so we enter that by 'f(2,3,:)=2;' for the last function.

Two conditions can also be entered this way for instance $f(2,1,0)=2$ and $f(2,1,-1)=4$ are entered by 'f(2,3,2)=2; f(2,3,1)=4;' for the last function.

Each time we change the boundary condition, we encounter new situation that is why enough attention will be paid by addressing more boundary conditions in following sections.

The sixth input argument of lap3d is user-supplied mean square error (section 2.5) for instance 10^{-5} is coded as 1e-5. The smaller is the error, the better is the solution for f.

The seventh input argument of lap3d is user-supplied maximum allowable iteration number which is an integer for instance 20.

The lap3d will terminate itself whichever between mean square error and maximum iteration number occurs first.

Also we assumed that all boundary values are 0 initially for f. The return of lap3d is a three dimensional array which is the solution for $f(x,y,z)$. The return from lap3d follows the coordinate convention of figure 2.3(a) for x, y, and z.

In summary the syntax of lap3d is user-supplied return variable for $f[m,n,k]$ or $f(x,y,z)$=lap3d(domain bounds as a six element row matrix, x resolution, y resolution, z resolution, boundary conditions as a string, user-decided mean square error, user-decided maximum iteration number).

Attention should be paid for the coordinate convention because MATLAB's coordinate differs from the textbook found one specially the definition of origin for a three dimensional array which is regularly exercised in 3D Laplace equation using finite difference.

The reader can get the function file through the email link of page ii.

3.3 Solution of a single impulse on boundary

Last two sections are prerequisite for this section. In this section we assume a single impulse function b_o $\delta(x_0,y_0,z_0)$ is located at any point on the boundary of the three dimensional solution space where b_o is the strength of the impulse and (x_0, y_0, z_0) is the location of the impulse. Following examples are demonstrated in this regard.

✦ Example 1

Solve Laplace equation $\nabla^2 f = 0$ using finite difference subject to $f(0,0,0) = 10V$ over $0 \le x < 15cm$, $0 < y < 10cm$, and $0 \le z \le 8cm$ where the spatial resolutions are $\Delta x = 5\ cm$, $\Delta y = 2.5\ cm$, and $\Delta z = 2\ cm$.

Solution:

The point $(x_0, y_0, z_0) = (0,0,0)$ does not include the finite difference lattice of section 3.1 so the lap3d function will not be useful. Even if the impulse were $f(x_0,0,0) = 10V$, we would not get any solution for the same reason. Similarly $f(0, y_0, 0) = 10V$ or $f(0,0,z_0) = 10V$ i.e. if the impulse is located exactly on the edge of the rectangular box, the solution is not obtainable which is a limitation of the FD.

✦ Example 2

In example 1 now the single impulse is given by $f(0,5cm,2cm) = 10V$ while other functional descriptions are the same.

Solution:

The point $(x_0, y_0, z_0) = (0,5cm,2cm)$ becomes $(1,3,2)$ in integer index domain subject to the resolution hence choose mean square error 10^{-10} and maximum iteration number 25 whichever is earlier. The boundary condition is inserted by 'f(1,3,2)=10;' so call the solver lap3d as:

```
>>f=lap3d([0 15 0 10 0 8],5,2.5,2,'f(1,3,2)=10;',1e-10,25) ↵
```

f(:,:,1) =

0	0	0	0
0	0	0	0
0	0	0	0
0	0	0	0
0	0	0	0

f(:,:,2) =

0	0	0	0
0	0.1193	0.0163	0
10.0000	0.5439	0.0380	0
0	0.1193	0.0163	0
0	0	0	0

f(:,:,3) =

0	0	0	0
0	0.0788	0.0154	0
0	0.1993	0.0288	0
0	0.0788	0.0155	0
0	0	0	0

f(:,:,4) =

0	0	0	0
0	0.0344	0.0083	0
0	0.0682	0.0140	0

$$\begin{array}{cccc} 0 & 0.0344 & 0.0083 & 0 \\ 0 & 0 & 0 & 0 \end{array}$$

f(:,:,5) =

$$\begin{array}{cccc} 0 & 0 & 0 & 0 \\ 0 & 0 & 0 & 0 \\ 0 & 0 & 0 & 0 \\ 0 & 0 & 0 & 0 \\ 0 & 0 & 0 & 0 \end{array}$$

Now we have the solution because $f(0,5cm,2cm)=10V$ is within the FD lattice. Clearly in above return the f(:,:,1) indicates solution for $f(x,y,0cm)$, f(:,:,2) indicates solution for $f(x,y,2cm)$, and so forth.

3.4 Solution of a linear impulse on boundary

Sections 3.1-3.3 are the prerequisite for this section. The linear impulse can be represented by $b_o\,\delta(x,y_0,z_0)$, $b_o\,\delta(x_0,y,z_0)$, or $b_o\,\delta(x_0,y_0,z)$ where b_o is the constant strength of the linear impulse. As quoted in section 3.3 if the linear impulse does not fall within the finite difference lattice, we do not get any solution. In the following we illustrate some problems regarding linear impulse.

◆ **Example 1**

Solve the Laplace equation $\nabla^2 f = 0$ using finite difference subject to $f(x,0,0)=10V$ over $0 \le x \le 15cm$, $0 \le y \le 10cm$, and $0 \le z \le 8cm$ where the spatial resolutions are $\Delta x = 5\ cm$, $\Delta y = 2.5\ cm$, and $\Delta z = 2\ cm$.

Solution:

Although the impulse extends along $0 \le x \le 15cm$, that does not belong to the finite difference lattice so the lap3d does not return any solution. Similarly we will not obtain any solution for $f(0,y,0)=10V$, nor for $f(0,0,z)=10V$. This is even true on other edges for instance $f(0,y,8cm)=10V$ or $f(15cm,y,8cm)=10V$.

◆ **Example 2**

In example 1 now we have the boundary condition $f(5cm,y,8cm)=10V$ while other functional descriptions are the same.

Solution:

In integer index domain the $(5cm,y,8cm)$ translates to $(2,n,5)$. Knowing so the boundary value is inserted by 'f(2,:,5)=10;'. Choose mean square error 10^{-10} and maximum iteration number 25 whichever is earlier hence call the solver as:

>>f=lap3d([0 15 0 10 0 8],5,2.5,2,'f(2,:,5)=10;',1e-10,25) ↵

f(:,:,1) =

```
              0   0   0   0
              0   0   0   0
              0   0   0   0
              0   0   0   0
              0   0   0   0
f(:,:,2) =
              0      0        0      0
              0   0.6413   0.1394    0
              0   0.8563   0.1913    0
              0   0.6414   0.1394    0
              0      0        0      0
f(:,:,3) =
              0      0        0      0
              0   1.7386   0.2767    0
              0   2.2313   0.3734    0
              0   1.7386   0.2767    0
              0      0        0      0
f(:,:,4) =
              0      0        0      0
              0   4.1453   0.3397    0
              0   4.8911   0.4419    0
              0   4.1453   0.3397    0
              0      0        0      0
f(:,:,5) =
              0   10   0   0
              0   10   0   0
              0   10   0   0
              0   10   0   0
              0   10   0   0
```

The return f(:,:,1) indicates solution for $f(x,y,0)$, f(:,:,2) for $f(x,y,2cm)$, and so on.

3.5 Solution of a planar impulse on boundary

We define a planar impulse by $b_o\,\delta(x,y,z_0)$, $b_o\,\delta(x_0,y,z)$, or $b_o\,\delta(x,y_0,z)$ where b_o is the constant strength of the planar impulse. Unlike the single or linear counterpart the lap3d always returns solution for a planar impulse which is demonstrated in the following example. If any condition is required for the solution, that is taken from previous sections.

◆ **Example 1**

Solve the Laplace equation $\nabla^2 f = 0$ using finite difference subject to $f(x,y,0) = 10V$ over the three dimensional domain and resolutions of previous examples.

Solution:

In integer index domain the $(x,y,0)$ translates to $(m,n,1)$. Knowing so the boundary value is inserted by 'f(:,:,1)=10;'. Choose mean square error and maximum iteration number as conducted in previous sections:

```
>>f=lap3d([0 15 0 10 0 8],5,2.5,2,'f(:,:,1)=10;',1e-10,25) ↵
```

f(:,:,1) =

```
            10   10   10   10
            10   10   10   10
            10   10   10   10
            10   10   10   10
            10   10   10   10
```

f(:,:,2) =

```
    0        0        0      0
    0     4.4850   4.4850    0
    0     5.3330   5.3330    0
    0     4.4850   4.4850    0
    0        0        0      0
```

f(:,:,3) =

```
    0        0        0      0
    0     2.0153   2.0153    0
    0     2.6047   2.6047    0
    0     2.0153   2.0154    0
    0        0        0      0
```

f(:,:,4) =

```
    0        0        0      0
    0     0.7808   0.7808    0
    0     1.0477   1.0477    0
    0     0.7808   0.7808    0
    0        0        0      0
```

f(:,:,5) =

```
    0   0   0   0
    0   0   0   0
    0   0   0   0
    0   0   0   0
    0   0   0   0
```

In above f(:,:,1), f(:,:,2), etc have earlier quoted meanings. The $f(x,y,0)=10V$ refers to $z=0$ that is why contents of f(:,:,1) are all 10.

As another example if we had $f(x,10cm,z)=10V$ any discrete point of which is (m,5,k), the command would have been:

>>f=lap3d([0 15 0 10 0 8],5,2.5,2,'f(:,5,:)=10;',1e-10,25); ↵

Again if we had $f(15cm,y,z)=10V$ any discrete point of which is (4,n,k), the command would have been:

>>f=lap3d([0 15 0 10 0 8],5,2.5,2,'f(4,:,:)=10;',1e-10,25); ↵

◆ Example 2

Solve the Laplace equation $\nabla^2 f=0$ using finite difference subject to the planar segment boundary condition $f(x,y,8cm)=10V$ over the three dimensional domain and resolutions of ongoing example with $10cm \le x \le 15cm$ and $2.5cm \le y \le 10cm$.

Solution:

In this problem the boundary condition spans over a portion on the $z=8cm$ plane. The interval $10cm \le x \le 15cm$ translates to $m=3$ through $m=4$ and $2.5cm \le y \le 10cm$ to $n=2$ through $n=5$. Not to mention $z=8cm$ translates to $k=5$. As a string the boundary code is **'f(3:4,2:5,5)=10;'**. Just call the solver with relevant conditions similar to previous examples:

```
>>f=lap3d([0 15 0 10 0 8],5,2.5,2,'f(3:4,2:5,5)=10;',1e-10,25) ↵
```

f(:,:,1) =

0	0	0	0
0	0	0	0
0	0	0	0
0	0	0	0
0	0	0	0

f(:,:,2) =

0	0	0	0
0	0.1393	0.6413	0
0	0.1913	0.8563	0
0	0.1394	0.6414	0
0	0	0	0

f(:,:,3) =

0	0	0	0
0	0.2767	1.7386	0
0	0.3734	2.2313	0
0	0.2767	1.7386	0
0	0	0	0

f(:,:,4) =

0	0	0	0
0	0.3396	4.1453	0
0	0.4419	4.8911	0
0	0.3397	4.1453	0
0	0	0	0

f(:,:,5) =

0	0	0	0
0	0	10	10
0	0	10	10
0	0	10	10
0	0	10	10

If you notice, the boundary condition is visible in **f(:,:,5)** or at $z = 8cm$ plane. The $f(x, y, 8cm) = 10V$ over $10cm \leq x \leq 15cm$ and $2.5cm \leq y \leq 10cm$ is occupied as

$$\begin{bmatrix} 10 & 10 \\ 10 & 10 \\ 10 & 10 \\ 10 & 10 \end{bmatrix}$$ i.e. a planar segment not the whole plane.

Example of another planar segment boundary condition is $f(15cm, y, z) = 10V$ over $5cm \leq y \leq 10cm$ and $4cm \leq z \leq 8cm$ which would require the following:

```
>>f=lap3d([0 15 0 10 0 8],5,2.5,2,'f(4,3:5,3:5)=10;',1e-10,25); ↵
```

Because in index domain $5cm \leq y \leq 10cm$ and $4cm \leq z \leq 8cm$ translate to $n=3$ to 5 and $k=3$ to 5 respectively while $x = 15cm$ indicates $m=4$.

3.6 Solution of a functional impulse on boundary

We define a functional impulse by $g(x,y)$, $g(y,z)$, or $g(z,x)$ while the third coordinate point is constant for instance the third coordinate z is constant in $g(x,y)$ and the $g(x,y)$ is some function of x and y. In section 3.2

we did not address other aspect of lap3d which is its accessibility to numerical boundary condition rather string one.

Recall the fifth input argument of lap3d which we enter as a string for instance by writing 'f(4,3:5,3:5)=10;'. Now the argument will be a matrix representing $g(x,y)$, $g(y,z)$, or $g(z,x)$ for a functional impulse. The consequential query is how to calculate the 2D function g ? OK, first adopt the section 2.3 illustrated grid point tactic but you exercise it at the command prompt or in script file and then pass the g samples to lap3d.

Owing to the coordinate convention difference we need to transpose g samples while argumenting into lap3d but only for $g(x,y)$ and $g(y,z)$. If it is $g(z,x)$, there is no need for transposition.

Another important point is the lap3d will ask you the third plane in integer index form which you input without the quote at the command prompt. Let us check the following examples in this regard.

◆ Example 1

Solve the Laplace equation $\nabla^2 f = 0$ using finite difference subject to $f(x,y,0) = 10(x+y)$ over the three dimensional domain and resolutions of previous examples.

Solution:

The x and y intervals are $0 \le x \le 15cm$ and $0 \le y \le 10cm$ with $\Delta x = 5\ cm$ and $\Delta y = 2.5\ cm$ respectively therefore the two variational row vectors are generated by:

```
>>x=0:5:15; y=0:2.5:10; ↵
```

From the problem description we discover $g(x,y) = 10(x+y)$ and $z = 0$. The $g(x,y)$ is computed by:

```
>>[X,Y]=meshgrid(x,y); g=10*(X+Y); ↵
```

In above the g holds $g(x,y)$ samples. The $z = 0$ translates to $k = 1$ in integer domain so we should write the $z = 0$ plane as f(:,:,1) which will be entered from the command prompt once the lap3d has been run. Not to mention g should be entered as g' for the coordinate convention reason. Having all input arguments available and drawing previous solution constraints, call the solver as:

```
>>f=lap3d([0 15 0 10 0 8],5,2.5,2,g',1e-10,25); ↵
Enter the boundary condition plane: f(:,:,1) ↵
>>
```

Above execution indicates that the solution is found and available in f. Call the f to see its contents:

```
>>f ↵
```

f(:,:,1) =

    0   50 100 150

```
25   75  125  175
50  100  150  200
75  125  175  225
100  150  200  250
```

f(:,:,2) =

```
0      0        0      0
0  38.5422  57.5706   0
0  55.5397  77.7859   0
0  54.5551  73.5836   0
0      0        0      0
```

f(:,:,3) =

```
0      0        0      0
0  18.9951  26.3045   0
0  27.9151  37.2043   0
0  24.0797  31.3892   0
0      0        0      0
```

f(:,:,4) =

```
0      0        0      0
0   7.7658  10.2758   0
0  11.4341  14.7591   0
0   9.2439  11.7539   0
0      0        0      0
```

f(:,:,5) =

```
0  0  0  0
0  0  0  0
0  0  0  0
0  0  0  0
0  0  0  0
```

Note that $f(x,y,0) = 10(x+y)$ is mapped on plane f(:,:,1) that is what our boundary condition is unlike the constant counterpart of previous sections.

◆ Example 2

In example 1 we have the boundary condition given as $f(0,y,z)$ $= \dfrac{(y+z)^2}{120}$.

Solution:

Now $g(y,z)$ is given and the computing requires the following:

```
>>y=0:2.5:10; z=0:2:8; ↵
>>[Y,Z]=meshgrid(y,z); g=(Y+Z).^2/120; ↵
>>f=lap3d([0 15 0 10 0 8],5,2.5,2,g',1e-10,25); ↵
Enter the boundary condition plane: f(1,:,:) ↵
>>f ↵
```

f(:,:,1) =

```
0       0  0  0
0.0521  0  0  0
0.2083  0  0  0
0.4688  0  0  0
0.8333  0  0  0
```

f(:,:,2) =

0.0333	0	0	0
0.1688	0.0388	0.0062	0
0.4083	0.0722	0.0104	0
0.7521	0.0820	0.0095	0
1.2000	0	0	0

f(:,:,3) =

0.1333	0	0	0
0.3521	0.0655	0.0096	0
0.6750	0.1158	0.0158	0
1.1021	0.1272	0.0144	0
1.6333	0	0	0

f(:,:,4) =

0.3000	0	0	0
0.6021	0.0650	0.0078	0
1.0083	0.1110	0.0128	0
1.5188	0.1231	0.0117	0
2.1333	0	0	0

f(:,:,5) =

0.5333	0	0	0
0.9187	0	0	0
1.4083	0	0	0
2.0021	0	0	0
2.7000	0	0	0

Where do you think $f(0,y,z)$ is mapped onto? The answer is in the first column of every page.

◆ Example 3

In example 1 we have the boundary condition given as $f(x,10cm,z)$ $= e^{-x/7-z/8}$.

Solution:

Again $g(z,x)$ is given and there is no need to transpose its samples. The $y = 10cm$ is basically $n = 5$ so execution is the following:

```
>>z=0:2:8; x=0:5:15; ↵
>>[Z,X]=meshgrid(z,x); g=exp(-X/7-Z/8); ↵
>>f=lap3d([0 15 0 10 0 8],5,2.5,2,g,1e-10,25); ↵
Enter the boundary condition plane: f(:,5,:) ↵
>>f ↵
```

f(:,:,1) =

0	0	0	0
0	0	0	0
0	0	0	0
0	0	0	0
1.0000	0.4895	0.2397	0.1173

f(:,:,2) =

0	0	0	0
0	0.0097	0.0063	0
0	0.0332	0.0201	0
0	0.1083	0.0594	0
0.7788	0.3813	0.1866	0.0914

f(:,:,3) =

0	0	0	0
0	0.0127	0.0084	0

```
                0   0.0408   0.0251      0
                0   0.1152   0.0643      0
           0.6065   0.2969   0.1454   0.0712
```

f(:,:,4) =

```
                0        0        0        0
                0   0.0087   0.0058        0
                0   0.0279   0.0172        0
                0   0.0801   0.0446        0
           0.4724   0.2312   0.1132   0.0554
```

f(:,:,5) =

```
                0        0        0        0
                0        0        0        0
                0        0        0        0
                0        0        0        0
           0.3679   0.1801   0.0882   0.0432
```

Where do you perceive the boundary condition? The answer is as the last row of every page.

✦ Example 4

What if we have functional variation over a segment for instance in example 3 over $5cm \leq x \leq 10cm$ and $2cm \leq z \leq 6cm$?

Solution:

In this sort of problems padding by 0 is required beyond the segment. We get the samples of g first. Instead of the whole interval perform the computing only over the given segment:

```
>>z=2:2:6; x=5:5:10; [Z,X]=meshgrid(z,x); g=exp(-X/7-Z/8); ↵
```

Boundary functional values of the sector are in **g** and have them called by:

```
>>g ↵
```

g =

```
           0.3813   0.2969   0.2312
           0.1866   0.1454   0.1132
```

Had you called the **g** of example 3, you would have seen the following as its contents:

```
           1.0000   0.7788   0.6065   0.4724   0.3679
           0.4895   0.3813   0.2969   0.2312   0.1801
           0.2397   0.1866   0.1454   0.1132   0.0882
           0.1173   0.0914   0.0712   0.0554   0.0432
```

How to turn the size of this example **g** to the one in example 3? We need to pad 0 of size 1×1 along the two extreme diagonal points of **g** which is accomplished by the **padarray** of appendix B.8:

```
>>g=padarray(g,[1 1],0,'both') ↵
```

g =

```
           0        0        0        0      0
           0   0.3813   0.2969   0.2312      0
           0   0.1866   0.1454   0.1132      0
           0        0        0        0      0
```

In order to maintain identical symbology we assigned the return from the **padarray** to **g** again. Despite padding when you call the solver for the solution, full interval should be engaged so repetition of example 3 reveals the following:

```
>>f=lap3d([0 15 0 10 0 8],5,2.5,2,g,1e-10,25); ↵
Enter the boundary condition plane: f(:,5,:) ↵
>>f ↵
```

f(:,:,1) =

0	0	0	0
0	0	0	0
0	0	0	0
0	0	0	0
0	0	0	0

f(:,:,2) =

0	0	0	0
0	0.0097	0.0063	0
0	0.0332	0.0201	0
0	0.1083	0.0594	0
0	0.3813	0.1866	0

f(:,:,3) =

0	0	0	0
0	0.0127	0.0084	0
0	0.0408	0.0251	0
0	0.1152	0.0643	0
0	0.2969	0.1454	0

f(:,:,4) =

0	0	0	0
0	0.0087	0.0058	0
0	0.0279	0.0172	0
0	0.0801	0.0446	0
0	0.2312	0.1132	0

f(:,:,5) =

0	0	0	0
0	0	0	0
0	0	0	0
0	0	0	0
0	0	0	0

Where do you think the boundary condition spans around? The answer is on the pages f(:,:,2), f(:,:,3), and f(:,:,4) intersecting second and third columns in every last row.

We terminate the chapter upon the execution of this example.

Exercises

1. Solve Laplace equation $\nabla^2 f = 0$ using finite difference subject to single impulse $f(0cm,-1cm,1cm) = 30V$ over $0 \le x \le 8cm$, $-2cm \le y \le 0cm$, and $0 \le z \le 4cm$ where the spatial resolutions are $\Delta x = 2\ cm$, $\Delta y = 0.5\ cm$, and $\Delta z = 1\ cm$. Consider the mean square error 10^{-10} and maximum iteration number 25 whichever is earlier.

2. In problem 1 now solve for the linear impulse $f(0cm, y, 1cm) = 30V$.

3. In problem 1 now solve for the planar impulse $f(8cm, y, z) = 30V$.

4. In problem 1 now solve the equation for segmentary boundary condition $f(x, y, 4cm) = 30V$ where the segment is defined by $4cm \le x \le 6cm$ and $-1cm \le y \le 0cm$.

5. In problem 1 now solve the equation for segmentary boundary condition $f(x, -2cm, z) = 30V$ where the segment is defined by $4cm \le x \le 6cm$ and $2cm \le z \le 4cm$.

6. In problem 1 now solve for the functional impulse $g(x, y, 0) = 3x - 2y$.

7. In problem 1 now solve for the functional impulse $g(x, y, 0) = 3x - 2y$ but over the segment $4cm \le x \le 6cm$ and $-1cm \le y \le 0cm$.

8. In problem 1 now solve for the functional impulse $g(x, -2cm, z) = e^{-0.3z} + \cos x$.

9. In problem 1 now solve for the functional impulse $g(x, -2cm, z) = e^{-0.3z} + \cos x$ but over the segment $4cm \le x \le 6cm$ and $2cm \le z \le 4cm$.

Mohammad Nuruzzaman

Answers:

(1)

$$f(x,y,0cm) = \begin{bmatrix} 0 & 0 & 0 & 0 & 0 \\ 0 & 0 & 0 & 0 & 0 \\ 0 & 0 & 0 & 0 & 0 \\ 0 & 0 & 0 & 0 & 0 \\ 0 & 0 & 0 & 0 & 0 \end{bmatrix}$$

$$f(x,y,1cm) = \begin{bmatrix} 0 & 0 & 0 & 0 & 0 \\ 0 & 0.4101 & 0.0301 & 0.0019 & 0 \\ 30.0000 & 1.0461 & 0.0498 & 0.0028 & 0 \\ 0 & 0.4101 & 0.0301 & 0.0019 & 0 \\ 0 & 0 & 0 & 0 & 0 \end{bmatrix}$$

$$f(x,y,2cm) = \begin{bmatrix} 0 & 0 & 0 & 0 & 0 \\ 0 & 0.1145 & 0.0138 & 0.0012 & 0 \\ 0 & 0.1909 & 0.0204 & 0.0017 & 0 \\ 0 & 0.1145 & 0.0138 & 0.0012 & 0 \\ 0 & 0 & 0 & 0 & 0 \end{bmatrix}$$

$$f(x,y,3cm) = \begin{bmatrix} 0 & 0 & 0 & 0 & 0 \\ 0 & 0.0253 & 0.0042 & 0.0005 & 0 \\ 0 & 0.0376 & 0.0061 & 0.0007 & 0 \\ 0 & 0.0253 & 0.0042 & 0.0005 & 0 \\ 0 & 0 & 0 & 0 & 0 \end{bmatrix}$$

$$f(x,y,4cm) = \begin{bmatrix} 0 & 0 & 0 & 0 & 0 \\ 0 & 0 & 0 & 0 & 0 \\ 0 & 0 & 0 & 0 & 0 \\ 0 & 0 & 0 & 0 & 0 \\ 0 & 0 & 0 & 0 & 0 \end{bmatrix}$$

Hint: section 3.3

(2)

$$f(x,y,0cm) = \begin{bmatrix} 0 & 0 & 0 & 0 & 0 \\ 0 & 0 & 0 & 0 & 0 \\ 0 & 0 & 0 & 0 & 0 \\ 0 & 0 & 0 & 0 & 0 \\ 0 & 0 & 0 & 0 & 0 \end{bmatrix}$$

$$f(x,y,1cm) = \begin{bmatrix} 30.0000 & 0 & 0 & 0 & 0 \\ 30.0000 & 1.4563 & 0.0800 & 0.0047 & 0 \\ 30.0000 & 1.8664 & 0.1101 & 0.0066 & 0 \\ 30.0000 & 1.4563 & 0.0800 & 0.0047 & 0 \\ 30.0000 & 0 & 0 & 0 & 0 \end{bmatrix}$$

$$f(x,y,2cm) = \begin{bmatrix} 0 & 0 & 0 & 0 & 0 \\ 0 & 0.3055 & 0.0342 & 0.0030 & 0 \\ 0 & 0.4201 & 0.0480 & 0.0042 & 0 \\ 0 & 0.3055 & 0.0343 & 0.0030 & 0 \\ 0 & 0 & 0 & 0 & 0 \end{bmatrix}$$

$$f(x,y,3cm) = \begin{bmatrix} 0 & 0 & 0 & 0 & 0 \\ 0 & 0.0630 & 0.0103 & 0.0011 & 0 \\ 0 & 0.0883 & 0.0146 & 0.0016 & 0 \\ 0 & 0.0630 & 0.0103 & 0.0011 & 0 \\ 0 & 0 & 0 & 0 & 0 \end{bmatrix}$$

$$f(x,y,4cm) = \begin{bmatrix} 0 & 0 & 0 & 0 & 0 \\ 0 & 0 & 0 & 0 & 0 \\ 0 & 0 & 0 & 0 & 0 \\ 0 & 0 & 0 & 0 & 0 \\ 0 & 0 & 0 & 0 & 0 \end{bmatrix}$$

Hint: section 3.4

(3)

$$f(x,y,0cm) = \begin{bmatrix} 0 & 0 & 0 & 0 & 30 \\ 0 & 0 & 0 & 0 & 30 \\ 0 & 0 & 0 & 0 & 30 \\ 0 & 0 & 0 & 0 & 30 \\ 0 & 0 & 0 & 0 & 30 \end{bmatrix}$$

$$f(x,y,1cm) = \begin{bmatrix} 0 & 0 & 0 & 0 & 30.0000 \\ 0 & 0.0088 & 0.1246 & 1.8249 & 30.0000 \\ 0 & 0.0124 & 0.1728 & 2.3750 & 30.0000 \\ 0 & 0.0088 & 0.1246 & 1.8249 & 30.0000 \\ 0 & 0 & 0 & 0 & 30.0000 \end{bmatrix}$$

$$f(x,y,2cm) = \begin{bmatrix} 0 & 0 & 0 & 0 & 30.0000 \\ 0 & 0.0118 & 0.1589 & 2.1305 & 30.0000 \\ 0 & 0.0166 & 0.2209 & 2.7951 & 30.0000 \\ 0 & 0.0118 & 0.1589 & 2.1305 & 30.0000 \\ 0 & 0 & 0 & 0 & 30.0000 \end{bmatrix}$$

$$f(x,y,3cm) = \begin{bmatrix} 0 & 0 & 0 & 0 & 30.0000 \\ 0 & 0.0088 & 0.1246 & 1.8249 & 30.0000 \\ 0 & 0.0124 & 0.1728 & 2.3750 & 30.0000 \\ 0 & 0.0088 & 0.1246 & 1.8249 & 30.0000 \\ 0 & 0 & 0 & 0 & 30.0000 \end{bmatrix}$$

$$f(x,y,4cm) = \begin{bmatrix} 0 & 0 & 0 & 0 & 30 \\ 0 & 0 & 0 & 0 & 30 \\ 0 & 0 & 0 & 0 & 30 \\ 0 & 0 & 0 & 0 & 30 \\ 0 & 0 & 0 & 0 & 30 \end{bmatrix}$$

Hint: section 3.5

(4)

$$f(x,y,0cm) = \begin{bmatrix} 0 & 0 & 0 & 0 & 0 \\ 0 & 0 & 0 & 0 & 0 \\ 0 & 0 & 0 & 0 & 0 \\ 0 & 0 & 0 & 0 & 0 \\ 0 & 0 & 0 & 0 & 0 \end{bmatrix}$$

$$f(x,y,1cm) = \begin{bmatrix} 0 & 0 & 0 & 0 & 0 \\ 0 & 0.0314 & 0.1949 & 0.1918 & 0 \\ 0 & 0.0460 & 0.2980 & 0.2935 & 0 \\ 0 & 0.0335 & 0.2234 & 0.2202 & 0 \\ 0 & 0 & 0 & 0 & 0 \end{bmatrix}$$

$$f(x,y,2cm) = \begin{bmatrix} 0 & 0 & 0 & 0 & 0 \\ 0 & 0.0976 & 0.7990 & 0.7909 & 0 \\ 0 & 0.1490 & 1.3712 & 1.3594 & 0 \\ 0 & 0.1115 & 1.0906 & 1.0820 & 0 \\ 0 & 0 & 0 & 0 & 0 \end{bmatrix}$$

$$f(x,y,3cm) = \begin{bmatrix} 0 & 0 & 0 & 0 & 0 \\ 0 & 0.1976 & 2.4880 & 2.4757 & 0 \\ 0 & 0.3388 & 6.1642 & 6.1454 & 0 \\ 0 & 0.2693 & 5.4450 & 5.4309 & 0 \\ 0 & 0 & 0 & 0 & 0 \end{bmatrix}$$

$$f(x,y,4cm) = \begin{bmatrix} 0 & 0 & 0 & 0 & 0 \\ 0 & 0 & 0 & 0 & 0 \\ 0 & 0 & 30 & 30 & 0 \\ 0 & 0 & 30 & 30 & 0 \\ 0 & 0 & 30 & 30 & 0 \end{bmatrix}$$

Hint: section 3.5

(5)

$$f(x,y,0cm) = \begin{bmatrix} 0 & 0 & 0 & 0 & 0 \\ 0 & 0 & 0 & 0 & 0 \\ 0 & 0 & 0 & 0 & 0 \\ 0 & 0 & 0 & 0 & 0 \\ 0 & 0 & 0 & 0 & 0 \end{bmatrix}$$

$$f(x,y,1cm) = \begin{bmatrix} 0 & 0 & 0 & 0 & 0 \\ 0 & 0.2637 & 2.7012 & 2.6807 & 0 \\ 0 & 0.3164 & 2.5552 & 2.5281 & 0 \\ 0 & 0.1999 & 1.4209 & 1.4027 & 0 \\ 0 & 0 & 0 & 0 & 0 \end{bmatrix}$$

$$f(x,y,2cm) = \begin{bmatrix} 0 & 0 & 30.0000 & 30.0000 & 0 \\ 0 & 0.8287 & 17.4061 & 17.3597 & 0 \\ 0 & 0.8289 & 9.6305 & 9.5729 & 0 \\ 0 & 0.4779 & 4.2978 & 4.2606 & 0 \\ 0 & 0 & 0 & 0 & 0 \end{bmatrix}$$

$$f(x,y,3cm) = \begin{bmatrix} 0 & 0 & 30.0000 & 30.0000 & 0 \\ 0 & 0.7705 & 16.9936 & 16.9528 & 0 \\ 0 & 0.7532 & 9.1491 & 9.0991 & 0 \\ 0 & 0.4277 & 3.9993 & 3.9673 & 0 \\ 0 & 0 & 0 & 0 & 0 \end{bmatrix}$$

$$f(x,y,4cm) = \begin{bmatrix} 0 & 0 & 30 & 30 & 0 \\ 0 & 0 & 0 & 0 & 0 \\ 0 & 0 & 0 & 0 & 0 \\ 0 & 0 & 0 & 0 & 0 \\ 0 & 0 & 0 & 0 & 0 \end{bmatrix}$$

Hint: section 3.5

(6)

$$f(x,y,0cm) = \begin{bmatrix} 4 & 10 & 16 & 22 & 28 \\ 3 & 9 & 15 & 21 & 27 \\ 2 & 8 & 14 & 20 & 26 \\ 1 & 7 & 13 & 19 & 25 \\ 0 & 6 & 12 & 18 & 24 \end{bmatrix}$$

$$f(x,y,1cm) = \begin{bmatrix} 0 & 0 & 0 & 0 & 0 \\ 0 & 1.8138 & 3.1268 & 4.1365 & 0 \\ 0 & 2.2141 & 3.9076 & 5.1900 & 0 \\ 0 & 1.6167 & 2.9251 & 3.9394 & 0 \\ 0 & 0 & 0 & 0 & 0 \end{bmatrix}$$

$$f(x,y,2cm) = \begin{bmatrix} 0 & 0 & 0 & 0 & 0 \\ 0 & 0.4075 & 0.7140 & 0.8917 & 0 \\ 0 & 0.5491 & 0.9716 & 1.2146 & 0 \\ 0 & 0.3881 & 0.6937 & 0.8723 & 0 \\ 0 & 0 & 0 & 0 & 0 \end{bmatrix}$$

$$f(x,y,3cm) = \begin{bmatrix} 0 & 0 & 0 & 0 & 0 \\ 0 & 0.0905 & 0.1594 & 0.1895 & 0 \\ 0 & 0.1258 & 0.2225 & 0.2647 & 0 \\ 0 & 0.0887 & 0.1574 & 0.1876 & 0 \\ 0 & 0 & 0 & 0 & 0 \end{bmatrix}$$

$$f(x,y,4cm) = \begin{bmatrix} 0 & 0 & 0 & 0 & 0 \\ 0 & 0 & 0 & 0 & 0 \\ 0 & 0 & 0 & 0 & 0 \\ 0 & 0 & 0 & 0 & 0 \\ 0 & 0 & 0 & 0 & 0 \end{bmatrix}$$

Hint: section 3.5

(7)

$$f(x,y,0cm) = \begin{bmatrix} 0 & 0 & 0 & 0 & 0 \\ 0 & 0 & 0 & 0 & 0 \\ 0 & 0 & 14 & 20 & 0 \\ 0 & 0 & 13 & 19 & 0 \\ 0 & 0 & 12 & 18 & 0 \end{bmatrix}$$

$$f(x,y,1cm) = \begin{bmatrix} 0 & 0 & 0 & 0 & 0 \\ 0 & 0.0923 & 1.1741 & 1.5896 & 0 \\ 0 & 0.1576 & 2.8816 & 3.9742 & 0 \\ 0 & 0.1238 & 2.4695 & 3.4472 & 0 \\ 0 & 0 & 0 & 0 & 0 \end{bmatrix}$$

$$f(x,y,2cm) = \begin{bmatrix} 0 & 0 & 0 & 0 & 0 \\ 0 & 0.0459 & 0.3814 & 0.5001 & 0 \\ 0 & 0.0699 & 0.6500 & 0.8617 & 0 \\ 0 & 0.0521 & 0.5105 & 0.6818 & 0 \\ 0 & 0 & 0 & 0 & 0 \end{bmatrix}$$

$$f(x,y,3cm) = \begin{bmatrix} 0 & 0 & 0 & 0 & 0 \\ 0 & 0.0149 & 0.0941 & 0.1197 & 0 \\ 0 & 0.0218 & 0.1434 & 0.1834 & 0 \\ 0 & 0.0158 & 0.1069 & 0.1374 & 0 \\ 0 & 0 & 0 & 0 & 0 \end{bmatrix}$$

$$f(x,y,4cm) = \begin{bmatrix} 0 & 0 & 0 & 0 & 0 \\ 0 & 0 & 0 & 0 & 0 \\ 0 & 0 & 0 & 0 & 0 \\ 0 & 0 & 0 & 0 & 0 \\ 0 & 0 & 0 & 0 & 0 \end{bmatrix}$$

Hint: sections 3.5 and 3.6, Two stage padding is required as follows:
g=padarray(g,[2 2],0,'pre'); g=padarray(g,[0 1],0,'post');

(8)

$$f(x,y,0cm) = \begin{bmatrix} 2.0000 & 0.5839 & 0.3464 & 1.9602 & 0.8545 \\ 0 & 0 & 0 & 0 & 0 \\ 0 & 0 & 0 & 0 & 0 \\ 0 & 0 & 0 & 0 & 0 \\ 0 & 0 & 0 & 0 & 0 \end{bmatrix}$$

$$f(x,y,1cm) = \begin{bmatrix} 1.7408 & 0.3247 & 0.0872 & 1.7010 & 0.5953 \\ 0 & 0.1650 & 0.0800 & 0.9234 & 0 \\ 0 & 0.0821 & 0.0569 & 0.4832 & 0 \\ 0 & 0.0342 & 0.0290 & 0.2077 & 0 \\ 0 & 0 & 0 & 0 & 0 \end{bmatrix}$$

$$f(x,y,2cm) = \begin{bmatrix} 1.5488 & 0.1327 & -0.1048 & 1.5090 & 0.4033 \\ 0 & 0.0860 & -0.0081 & 0.9392 & 0 \\ 0 & 0.0508 & 0.0206 & 0.5352 & 0 \\ 0 & 0.0235 & 0.0167 & 0.2411 & 0 \\ 0 & 0 & 0 & 0 & 0 \end{bmatrix}$$

$$f(x,y,3cm) = \begin{bmatrix} 1.4066 & -0.0096 & -0.2471 & 1.3667 & 0.2611 \\ 0 & 0.0058 & -0.0846 & 0.7642 & 0 \\ 0 & 0.0086 & -0.0211 & 0.4098 & 0 \\ 0 & 0.0055 & -0.0021 & 0.1790 & 0 \\ 0 & 0 & 0 & 0 & 0 \end{bmatrix}$$

$$f(x,y,4cm) = \begin{bmatrix} 1.3012 & -0.1150 & -0.3524 & 1.2614 & 0.1557 \\ 0 & 0 & 0 & 0 & 0 \\ 0 & 0 & 0 & 0 & 0 \\ 0 & 0 & 0 & 0 & 0 \\ 0 & 0 & 0 & 0 & 0 \end{bmatrix}$$

Hint: sections 3.5 and 3.6

(9)

$$f(x,y,0cm) = \begin{bmatrix} 0 & 0 & 0 & 0 & 0 \\ 0 & 0 & 0 & 0 & 0 \\ 0 & 0 & 0 & 0 & 0 \\ 0 & 0 & 0 & 0 & 0 \\ 0 & 0 & 0 & 0 & 0 \end{bmatrix}$$

$$f(x,y,1cm) = \begin{bmatrix} 0 & 0 & 0 & 0 & 0 \\ 0 & -0.0001 & 0.0016 & 0.1198 & 0 \\ 0 & -0.0000 & 0.0043 & 0.1093 & 0 \\ 0 & 0.0000 & 0.0034 & 0.0594 & 0 \\ 0 & 0 & 0 & 0 & 0 \end{bmatrix}$$

$$f(x,y,2cm) = \begin{bmatrix} 0 & 0 & -0.1048 & 1.5090 & 0 \\ 0 & -0.0014 & -0.0296 & 0.8203 & 0 \\ 0 & -0.0010 & -0.0020 & 0.4302 & 0 \\ 0 & -0.0004 & 0.0033 & 0.1853 & 0 \\ 0 & 0 & 0 & 0 & 0 \end{bmatrix}$$

$$f(x,y,3cm) = \begin{bmatrix} 0 & 0 & -0.2471 & 1.3667 & 0 \\ 0 & -0.0032 & -0.0899 & 0.7438 & 0 \\ 0 & -0.0024 & -0.0277 & 0.3864 & 0 \\ 0 & -0.0011 & -0.0063 & 0.1647 & 0 \\ 0 & 0 & 0 & 0 & 0 \end{bmatrix}$$

$$f(x,y,4cm) = \begin{bmatrix} 0 & 0 & -0.3524 & 1.2614 & 0 \\ 0 & 0 & 0 & 0 & 0 \\ 0 & 0 & 0 & 0 & 0 \\ 0 & 0 & 0 & 0 & 0 \\ 0 & 0 & 0 & 0 & 0 \end{bmatrix}$$

Hint: sections 3.5 and 3.6, Two stage padding is required as follows:
g=padarray(g,[2 2],'pre'); g=padarray(g,[1 0],'post');

Chapter 4

$$\nabla^2 f = 0$$

Visualization and Application on Electromagnetics

This chapter is important in the sense that whatever model, theory, or programming tactic addressed beforehand will be tested on electromagnetic system application. Since in-depth coverage of 3D space incorporating cylindrical and spherical systems is huge, we focus for the most part our discussion on Cartesian system. Despite this is not a text book, theoretical notion and symbology are put in perspective to connect the available or expected solution with ours by addressing the following conceptual sequence:

- ❖ ❖ Ways to visualize 3D FD solution - 1D/2D/3D context
- ❖ ❖ Handling electric potential and/or flux lines in 3D space
- ❖ ❖ Electromagnetic system property computing in 3D space
- ❖ ❖ Ways to handle composite electromagnetic system in 3D space

4.1 How to visualize the solution of Laplace equation?

In chapter 3 we addressed how to obtain numerical solution of Laplace equation using finite difference technique in three dimensional context. Straightforwardly the solutions of $f(x,y,z)$ subject to variety of boundary conditions are obtained and stored in a workspace variable f.

The visualization is user-decided and can be one, two, or three dimensional, few of which are addressed in the sequel:

(1) $f(x,y_0,z_0)$ versus x which is a one dimensional plot similarly $f(x_0,y,z_0)$ versus y or $f(x_0,y_0,z)$ versus z may also appear,

(2) $f(x,y,z_0)$ versus x and y which is a two dimensional plot similarly $f(x_0,y,z)$ versus y and z or $f(x,y_0,z)$ versus z and x may also appear, and

(3) $f(x,y,z)$ versus x, y, and z which is a three dimensional plot.

Given the nature of computer graphics, cases 1, 2, and 3 are handled by two, three, and four dimensional visualizations respectively. MATLAB embedded functions easily render the above visualizations which are exemplified in the following sections. Chapters 2 and 3 are the prerequisite of this chapter.

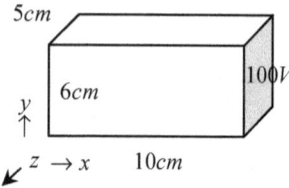

Figure 4.1(a) A rectangular box containing a potential on the right side

4.2 Two dimensional visualizations of Laplace equation solution

Figure 4.1(a) depicted rectangular box contains a potential $100\,V$ on the right side plane while the other sides are at $0\,V$. The box is formed by $0 \le x \le 10cm$, $0 \le y \le 6cm$, and $-5cm \le z \le 0$. Consider 21 points in each coordinate direction, determine the potential solution using 3D FD, and graph the following.

Problem 1: $f(x,3cm,-2cm)$ versus x

Solution:

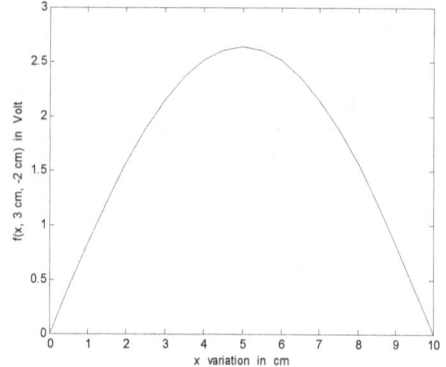

Figure 4.1(b) Plot of $f(x,3cm,-2cm)$ versus x

First determine the three resolutions Δx, Δy, and Δz:

```
>>dx=10/20; dy=6/20; dz=5/20; ↵          ← dx⇔ Δx , so as others
```

The nonzero potential is on $x=10cm$ plane so the boundary condition string is 'f(21,:,:)=100;'. Assume the mean square error 10^{-10} and maximum iteration number 25 and call the solver lap3d like chapter 3:

```
>>f=lap3d([0 10 0 6 -5 0],dx,dy,dz,'f(21,:,:)=100;',1e-10,25); ↵
```

The solution of $f(x,y,z)$ is available in **f**. The $f(x,3cm,-2cm)$ graph needs x sample generation which is:
>>x=0:dx:10; ⏎
Exercise appendix C explained plot for $f(x,3cm,-2cm)$ versus x as follows:
>>plot(x,f(:,11,13)) ⏎
In index domain $(x,3cm,-2cm)$ translates to $(m,11,13)$ for this reason f(:,11,13) is inserted into the plot. Figure 4.1(b) is the outcome from last execution.

Problem 2: $f(2cm,y,-2cm)$ versus y

Solution:
>>y=0:dy:6; plot(y,f(5,:,13)) ⏎
The $(2cm,y,-2cm)$ is equivalent to $(5,n,13)$. Graph is not shown for the space reason.

Problem 3: $f(3cm,3cm,z)$ versus z

Solution:
>>z=-5:dz:0; plot(z,squeeze(f(7,11,:))) ⏎
The $(3cm,3cm,z)$ is equivalent to $(7,11,k)$. While handling a linear data along the z direction of a 3D array for a fixed x and y, we get the array indexing as $(:,:,1)$, $(:,:,2)$, etc (like row or column linear data) in which the first two index elements are unwanted and in order to get rid of those we exercise the command **squeeze**. The graph is not shown for space reason.

Figure 4.2(a) Contour plot of $f(x,y,-2cm)$ versus x and y - right side figure

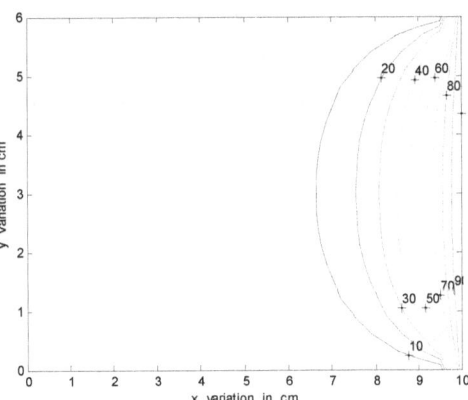

4.3 Three dimensional visualizations of Laplace equation solution

In this section we handle basically section 4.1 mentioned case 2. In three dimensional visualization there are several graph types, some of which are addressed in the sequel. Peripheral graphical properties for instance x or y axis labeling we are not going to address.

Contour plot:

A contour plot turns a three dimensional plot to two dimensional one in order to have 2D convenience and appendix C explained contour is invoked.

Problem 1:

Graph $f(x,y,-2cm)$ versus x and y as a contour plot for the section 4.2 mentioned rectangular box.

Solution:

The samples of $f(x,y,z)$ are available in f. Generate the x and y directed variations as a row matrix by:

```
>>x=0:dx:10; y=0:dy:6; ↵
```

The $(x,y,-2cm)$ is tantamount to $(m,n,13)$ in FD domain so get the planar data matrix by:

```
>>fxy=f(:,:,13); ↵
>>clabel(contour(x,y,fxy)) ↵
```

The above fxy is a user-chosen variable that holds $f(x,y,-2cm)$ samples as a rectangular matrix. Figure 4.2(a) depicts the contour plot which is also known as equipotential line in electromagnetics. The command clabel adds contour labeling once contour has been called.

Problem 2:

Graph the $f(5cm, y, z)$ versus y and z as a contour plot for the section 4.2 mentioned rectangular box.

Solution:

Analogous commands are the following:

```
>>dy=6/20; dz=5/20; y=0:dy:6; z=-5:dz:0; fyz=f(11,:,:); ↵
>>clabel(contour(y,z,squeeze(fyz))) ↵
```

The graph is similar to figure 4.2(a) with rotation by 90 degrees and not shown for space reason. See the last section for squeeze and we have $(5cm, y, z) \Leftrightarrow (11, n, k)$. The fyz is user-chosen that holds $f(5cm, y, z)$ samples as a rectangular matrix.

Problem 3:

Graph the $f(x, 4.5cm, z)$ versus z and x as a contour plot for the section 4.2 mentioned rectangular box.

Solution:

The commands you execute with ongoing symbology are the following:

```
>>dx=10/20; dz=5/20; z=-5:dz:0; x=0:dx:10;  ↵
>>fzx=f(:,16,:);  ↵
>>clabel(contour(z,x,squeeze(fzx)))  ↵
```

Here we have $(x, 4.5cm, z) \Leftrightarrow (m, 16, k)$ and the **squeeze** is required too. Figure 4.2(b) shows the equipotential line plot. User-chosen **fzx** holds $f(x, 4.5cm, z)$ samples as a rectangular matrix.

Surface plot:

Despite 3D to 2D visualization convenience by **contour**, a realistic 3D graph is often required. If a contour plot is available, so is the surface for the same $f(x, y, z)$ samples. The grapher we exercise is the **surf** of appendix C and call the surface plotter for every example of just conducted contour plot. In order to maintain continuity we may repeat previously conducted commands.

Problem 1:

Graph $f(x, y, -2cm)$ versus x and y as a surface plot for the section 4.2 mentioned rectangular box.

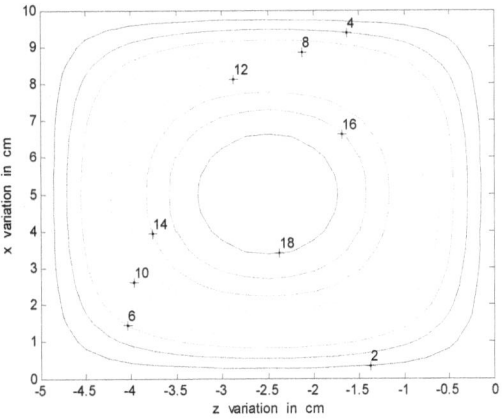

Figure 4.2(b)   Contour plot of $f(x, 4.5cm, z)$ versus z and x

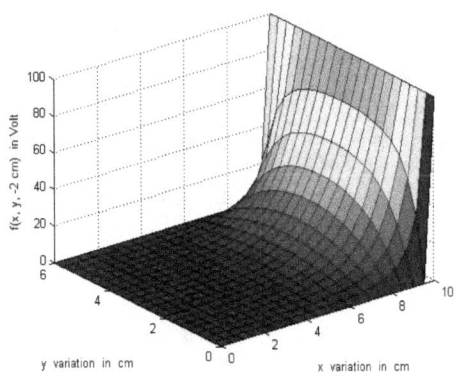

Figure 4.2(c)   Surface plot of $f(x, y, -2cm)$ versus x and y

Solution:

Execute the following commands:

```
>>x=0:dx:10; y=0:dy:6; ⏎
>>fxy=f(:,:,13); ⏎
>>surf(x,y,fxy) ⏎
```

As a result MATLAB prompted you with the surface graph of figure 4.2(c).

Problems 2 and 3:

Graph the $f(5cm, y, z)$ and $f(x, 4.5cm, z)$ as surface plots for the section 4.2 mentioned rectangular box.

Solution:

You need the commands surf(y,z,squeeze(fyz)) and surf(z,x,squeeze(fzx)) respectively with earlier symbol retainings and meanings. The surface graphs drawn this way may not appear smooth in color due to poor resolution. Better perceived surface plot is obtained by executing the following command:

```
>>shading interp ⏎
```

Digital image plot:

Laplace equation solution data for contour and surface plots can be maneuvered to view a digital image plot. The image plot is basically a normalized data plotter between 0 and 1. The 1 means white while 0 does black and intermediate values refer to gray. MATLAB image plotter is imshow which uses a syntax imshow(rectangular matrix representing the image where the elements are between 0 and 1).

Problem:

Graph the $f(x, y, -2cm)$ versus x and y as an image for the section 4.2 mentioned rectangular box.

↑
y

$x \rightarrow$

Figure 4.2(d) Image plot of $f(x, y, -2cm)$ versus x and y

Solution:

From the contour plot we know that fxy holds the $f(x, y, -2cm)$ samples as a rectangular matrix but not sure about the data range. Appendix B.5 mentioned max and min help us determine the range:

```
>>max(fxy(:)) ⏎

ans =
```
100

We first convert fxy as a column matrix by fxy(:) then exercise max over that. Anyhow the range of data in fxy is 0-100 hence dividing every element by 100 provides the intended range 0-1 and call the image grapher as:

>>imshow(fxy/100) ↵

Figure 4.2(d) shows the image plot. In the figure you find gray area close to the right side of the image meaning concentration of potential towards the right side. Most portion of the image is black meaning potential close to 0. The image displayed is in black and white form, color one would have been better. Unfortunately color image is defined better on $f[m,n]$ than $f(x,y)$ samples. There is an embedded function by the name im2uint8 which turns 0-1 values to unsigned 8 bit integer i.e. converts $f(x,y)$ samples to $f[m,n]$ in [0,255] scale. We may add artificial color for the color image display and execute the following:

>>imshow(im2uint8(fxy/100),cool) ↵

The imshow also accepts two input arguments, first and second of which are $f[m,n]$ matrix and colormap respectively which is what is executed above. There are many embedded artificial colormaps, one of which is cool and inserted as second input argument without quote. Other colormaps are hsv, summer, bone, prism, etc. Execute help colormap for more maps assistance. Sorry folks the text is in black and white form so we are unable to depict the color image certainly MATLAB will not disappoint you.

4.4 Electric field graphing on 3D FD

Once Laplace equation $\nabla^2 f(x,y,z) = 0$ has been solved in 3D space, the electric field $\bar{E}$ is given by $\bar{E}(x,y,z) = -\nabla f(x,y,z) = E_x(x,y,z)\bar{a}_x + E_y(x,y,z)\bar{a}_y + E_z(x,y,z)\bar{a}_z$, every element in the equation is in 3D space. The reader needs the basics of sections 2.7 and 2.8 for gradient computing. In this section our focus is mainly on graphical visualization. Whatever graph we exercised in previous sections can be exercised on $E_x(x,y,z)$, $E_y(x,y,z)$, and $E_z(x,y,z)$ but one at a time certainly as a scalar function despite $\bar{E}(x,y,z)$ being a vector.

Since $\bar{E}(x,y,z)$ is a vector, we had better graph the vector lines on the field so that flow direction is perceived. The embedded function streamslice displays the vector field lines with a syntax streamslice(only x point coordinates as a rectangular matrix, only y point coordinates as a rectangular matrix,[ ], only x component of the field as a rectangular matrix, only y component of the field as a rectangular matrix,[ ],0,0,[ ]). Why the input argument should be [ ] and 0 will be addressed later. We explained the syntax considering $E_x(x,y,z)\bar{a}_x + E_y(x,y,z)\bar{a}_y$ in fact $E_y(x,y,z)\bar{a}_y + E_z(x,y,z)\bar{a}_z$ or

$E_z(x,y,z)\overline{a}_z + E_x(x,y,z)\overline{a}_x$ can be graphed in a similar fashion too. Freeze one coordinate and graph the field lines of the other two e.g. $E_x(x,y,z_0)\overline{a}_x + E_y(x,y,z_0)\overline{a}_y$.

The problem is after discrete derivative has been found on $f(x,y,z)$, $\overline{E}(x,y,z)$ components evolve with different array sizes which create a problem for the vector field plotting because **streamslice** needs identical size grid point matrices. One way to solve the problem is repeat the last sample along particular direction for example along x. Let us go through the following examples.

◆ Example 1

Plot the vector field lines of electric field $E_x(x,y,-2cm)\overline{a}_x + E_y(x,y,-2cm)\overline{a}_y$ for section 4.2 cited rectangular box potential solution.

Solution:

We repeat the earlier execution for convenience:

```
>>dx=10/20; dy=6/20; dz=5/20; ↵
>>f=lap3d([0 10 0 6 -5 0],dx,dy,dz,'f(21,:,:)=100;',1e-10,25); ↵
```

Therefore potential $f(x,y,z)$ is available in **f** as a 3D array. The $E_x(x,y,z)$ and $E_y(x,y,z)$ are computed by (assigned to user-chosen variables **Gx** and **Gy** respectively):

```
>>Gx=-diff(f,1,2)/dx; Gy=-diff(f,1,1)/dy; ↵
```

If you look at the array sizes of **Gx** and **Gy** in workspace browser, you find them as 21×20×21 and 20×21×21 respectively that is how the complication starts for **streamslice**. Each should be of size 21×21×21 i.e. the size of $f(x,y,z)$. In every page of **Gx** we repeat the last column (i.e. the 20th) for **Gx** to get **Ex**:

```
for p=1:21
        Ex(:,:,p)=[Gx(:,:,p) Gx(:,20,p)];
end
```

The **p** in above is just a for loop counter (appendix B.4). The p-th page in **Gx** is **Gx(:,:,p)**. The **Gx(:,20,p)** indicates 20th column in the p-th page. The **Gx(:,:,p)** and **Gx(:,20,p)** are placed side by side (appendix B.3) by **[Gx(:,:,p) Gx(:,20,p)]** and put to user-chosen **Ex** but to the p-th page of **Ex** i.e. **Ex(:,:,p)=[Gx(:,:,p) Gx(:,20,p)];**.

Likewise in every page of **Gy** we repeat the last row (i.e. the 20th) for **Gy** to get **Ey**:

```
for p=1:21
        Ey(:,:,p)=[Gy(:,:,p);Gy(20,:,p)];
end
```

Now you find the sizes of **Ex** or **Ey** as 21×21×21 in workspace browser certainly indicative of $E_x(x,y,z)$ and $E_y(x,y,z)$ respectively. As the problem demands, $E_x(x,y,-2cm)\overline{a}_x + E_y(x,y,-2cm)\overline{a}_y$ is obtained from **Ex(:,:,13)** and **Ey(:,:,13)** subject to the given resolutions - each as a two dimensional array.

In order to fit into **streamslice**, we generate x and y directed variations as follows:

```
>>x=0:dx:10; y=0:dy:6; ⏎
```

From the last two row matrices, we generate **streamslice** required x only and y only rectangular grid point matrices by the **meshgrid** as follows:

```
>>[X,Y]=meshgrid(x,y); ⏎
```

Invoke the vector field plotter as:

```
>>streamslice(X,Y,[ ],Ex(:,:,13),Ey(:,:,13),[ ],0,0,[ ]); ⏎
```

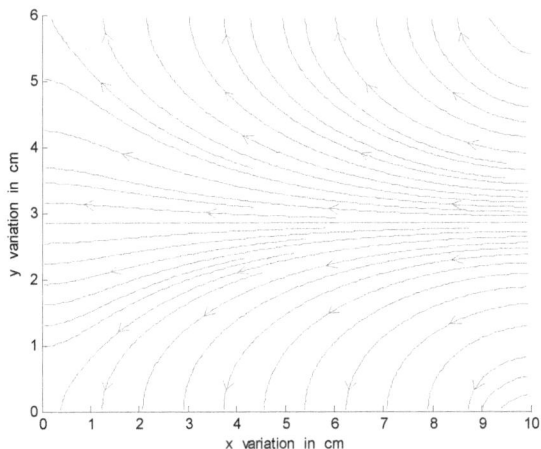

Figure 4.3(a) Vector electric field lines of
$E_x(x,y,-2cm)\overline{a}_x + E_y(x,y,-2cm)\overline{a}_y$ versus x and y

Figure 4.3(a) presents the vector electric field lines of $E_x(x,y,-2cm)\overline{a}_x + E_y(x,y,-2cm)\overline{a}_y$ versus x and y.

On another plane for instance $E_x(x,y,-4cm)\overline{a}_x + E_y(x,y,-4cm)\overline{a}_y$, you need the following command:

```
>>streamslice(X,Y,[ ],Ex(:,:,5),Ey(:,:,5),[ ],0,0,[ ]); ⏎
```

The graph is not shown for space reason, very similar to the above one.

◆ **Example 2**

Plot the vector field lines of electric field $E_x(x,0cm,z)\overline{a}_x + E_z(x,0cm,z)\overline{a}_z$ for section 4.2 cited rectangular box potential solution.

Solution:

From the example 1 we will be using **Ex**. For the $E_z(x,y,z)$ we execute the following:

```
>>Ez=-diff(f,1,3)/dz; ↵
```

Referring to the workspace browser of MATLAB, the size of **Ez** is 21×21×20. Now there is no need for the for loop along the z direction, we just repeat the last or 20th page and put the result to **Ez** again:

```
>>Ez(:,:,21)=Ez(:,:,20); ↵
```

After that z directed variation as a row matrix we get by:

```
>>z=-5:dz:0; ↵
```

For the sake of **streamslice** we generate the z only and x only grid points each as a rectangular matrix by:

```
>>[Z,X]=meshgrid(z,x); ↵
```

At last figure 4.3(b) depicted vector field lines we obtain by:

```
>>streamslice(Z,X,[ ],squeeze(Ez(:,1,:)),squeeze(Ex(:,1,:)),[ ],0,0,[ ]); ↵
```

See section 4.2 for **squeeze** and obviously **Ez(:,1,:)** and **Ex(:,1,:)** refer to $E_z(x,0cm,z)$ and $E_x(x,0cm,z)$ respectively.

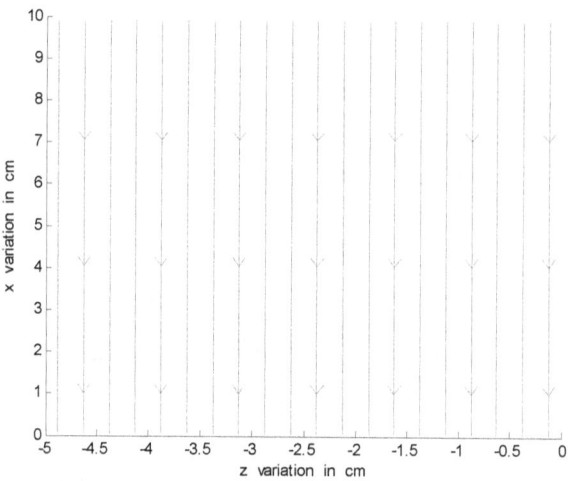

Figure 4.3(b) Vector electric field lines of
$E_x(x,0cm,z)\bar{a}_x + E_z(x,0cm,z)\bar{a}_z$ versus z and x

◆ **Example 3**

Plot the vector field lines of electric field $E_y(x,y,-4cm)\bar{a}_y + E_z(x,y,-4cm)\bar{a}_z$ for section 4.2 cited rectangular box potential solution.

Solution:

From the examples 1 and 2 we need and invoke **Ey** and **Ez** for $E_y(x,y,-4cm)$ and $E_z(x,y,-4cm)$ respectively. Generate the grid points by:

```
>>y=0:dy:6; z=-5:dz:0; [Y,Z]=meshgrid(y,z); ↵
>>streamslice(Y,Z,[ ],Ey(:,:,5),Ez(:,:,5),[ ],0,0,[ ]); ↵
```

The $z = -4cm$ plane turns to $k = 5$ in index domain i.e. $(x, y, -4cm) \Leftrightarrow [m, n, 5]$ anyhow the field plotter returns the figure 4.3(c) from above execution. Thus you may graph any planar electric fields.

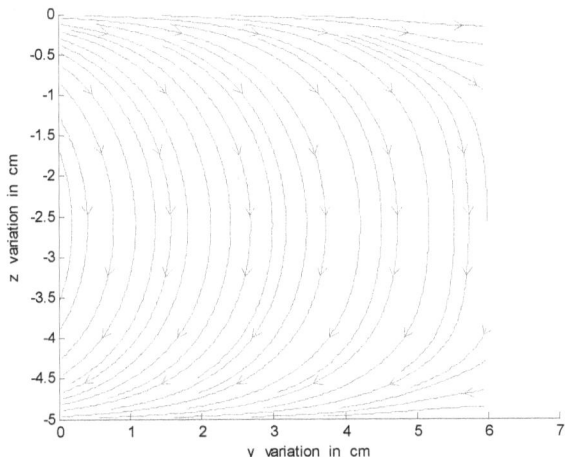

Figure 4.3(c) Vector electric field lines of $E_y(x, y, -4cm)\overline{a}_y + E_z(x, y, -4cm)\overline{a}_z$ against y and z

Note: It is important to point out that unit of $\overline{E}(x, y, z)$ in last examples is in V/cm.

4.5 Electric flux density computing and graphing

Electric flux density is given by $\overline{D} = \varepsilon_0 \overline{E}$ that is a scalar multiple of electric field $\overline{E}$ clearly in 3D space. As a function we have $\overline{D}(x, y, z) = D_x(x, y, z)\overline{a}_x + D_y(x, y, z)\overline{a}_y + D_z(x, y, z)\overline{a}_z$. Whatever graphing we conducted on $\overline{E}$ in section 4.4 can be exercised on $\overline{D}$ as a 3D array exactly in a similar fashion. The value of free space permittivity is $\varepsilon_0 = 8.854 \times 10^{-12} F/m$. We cite just one example on D computing and graphing in the sequel.

◆ **Example**

Plot the electric flux density lines (sometimes called flux lines) $D_x(x, y, -2cm)\overline{a}_x + D_y(x, y, -2cm)\overline{a}_y$ for section 4.2 cited rectangular box potential solution.

Solution:

Referring to example 1 of section 4.4, the $E_x(x, y, z)$ and $E_y(x, y, z)$ are available to variables Ex and Ey respectively. The electric flux density com-

-89-

ponents $D_x(x, y, z)$ and $D_y(x, y, z)$ we compute by:

>>eo=8.854e-12; Dx=eo*Ex*100; Dy=eo*Ey*100; ↵

In last command line **eo** stands for $\varepsilon_0 = 8.854 \times 10^{-12} F/m$ and user-chosen **Dx** and **Dy** hold $D_x(x, y, z)$ and $D_y(x, y, z)$ respectively. The multiplication by 100 is because of turning electric field from V/cm to V/m. For the electric flux line plot, just repeat the following but with the change from **E** to **D**:

>>x=0:dx:10; y=0:dy:6; [X,Y]=meshgrid(x,y); ↵
>>streamslice(X,Y,[ ],Dx(:,:,13),Dy(:,:,13),[ ],0,0,[ ]); ↵

The graph is similar to the figure 4.3(a) and not shown for space reason. Not to mention the unit of holdings in **Dx** or **Dy** is in C/m^2.

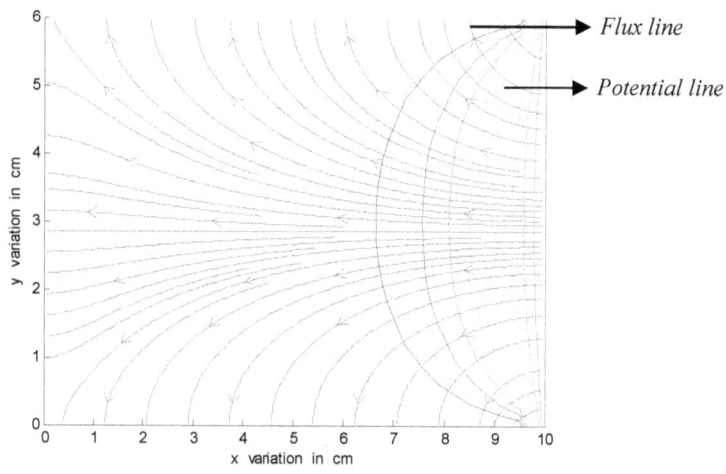

Figure 4.4(a) Electric flux and potential
lines plotted together

4.6 Visualization of electric potential and flux

It is possible to graph the electric potential and flux lines in a single plot. Sections 4.3 and 4.5 explain electric potential and flux line implementations respectively. Contour plot is basically electric potential lines. In order to obtain the single plot, first one of the two is graphed and then the second one is laid over by dint of command **hold**. Illustration is presented by one example.

◆ Example

Plot the electric flux lines $D_x(x, y, -2cm)\bar{a}_x + D_y(x, y, -2cm)\bar{a}_y$ and potential lines $f(x, y, -2cm)$ in a single plot for section 4.2 cited rectangular box potential solution.

Solution:

In last section you find the commands for electric flux lines $D_x(x,y,-2cm)\overline{a}_x + D_y(x,y,-2cm)\overline{a}_y$. Reexecute all commands and then exercise hold at the command prompt. Example 1 from contour plot of section 4.3 presents the electric potential line commands and reexecute those, you find the combined plot as in figure 4.4(a). Potential and flux lines are scalar and vector types respectively which are indicated in the figure too.

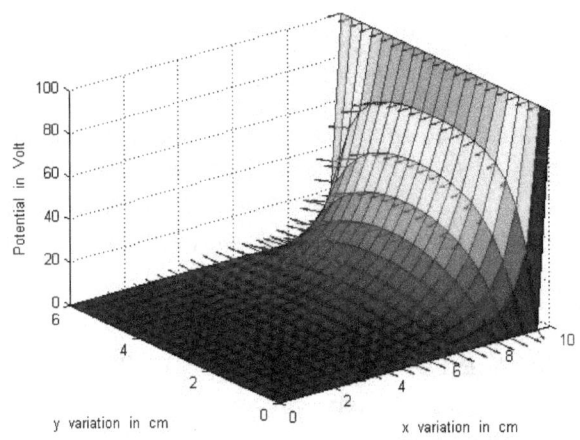

Figure 4.4(b) Surface vector $d\overline{S}$ of
$f(x,y,-2cm)$ against x and y

4.7 Surface normal of a potential function

Section 4.3 explained surface graph's valley is not so explicit on the graph. If normals are drawn on the surface, variation of the surface becomes more pronounced. In electromagnetics terminology this is basically elementary surface vector $d\overline{S}$.

In order to achieve this we invoke two embedded functions namely quiver3 and surfnorm. The surfnorm computes the normals through three output arguments i.e. [Nx,Ny,Nz]=surfnorm(X,Y,Z) where Nx, Ny, and Nz are the computed normals in the three directions and X and Y are the x only and y only grid point matrices in x and y directions respectively. The Z is basically functional value f(X,Y) at (X,Y), the X, Y, and Z are identical size rectangular matrices. The quiver3 just superimposes the normals on the drawn surface by surf which requires the syntax quiver3(X,Y,Z,Nx,Ny,Nz).

To talk about the order, first use surf for surface, then hold the surface by the command hold, next compute the normals by the surfnorm, and after that exercise the quiver3. One example is presented in this regard.

✦ Example

Plot $d\bar{S}$ of the potential surface $f(x,y,-2cm)$ for section 4.2 cited rectangular box solution.

Solution:

Reexecute the commands of sections 4.2 and 4.3 (problem 1 of surface plot) until you get the figure 4.2(c). After that execute the following maintaining the syntaxes of **surfnorm** and **quiver3**:

```
>>hold ⌋
>>[X,Y]=meshgrid(x,y); ⌋
>>[Nx,Ny,Nz]=surfnorm(X,Y,fxy); ⌋
>>quiver3(X,Y,fxy,Nx,Ny,Nz) ⌋
```

If the axes are set to minimum to maximum, the graph is perceived better which happens by the command **axis** with syntax **axis**([x -min $\quad$ x -max y -min $\quad$ y -max $\quad$ $f(x,y,-2cm)$ -min $\quad$ $f(x,y,-2cm)$ -max]) i.e. a six element row matrix and for the ongoing problem we do so by:

```
>>axis([0 10 0 6 0 100]) ⌋
```

The result is the figure 4.4(b) as expected.

4.8 Resistance computing using 3D FD

The fundamental idea concerning resistance computing in an electromagnetic system applies the expression $R = \dfrac{V_0}{\oint \sigma \bar{E} \circ d\bar{S}}$. In a given plate

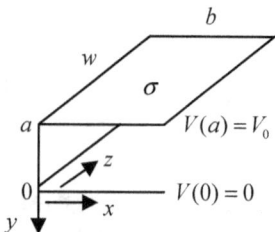

we apply a fixed potential V_0 which has conductivity σ (in mho/meter) and the $\bar{E}$ is the electric field given by $\bar{E} = -\nabla f(x,y,z)$ where $f(x,y,z)$ is the scalar potential function of coarse in 3D domain. The $d\bar{S}$

Figure 4.5(a) A 3D electromagnetic system with one constant boundary voltage along y direction

is the elementary surface area and the denominator integration evolves from closed integration. The $f(x,y,z)$ is obtained by solving Laplace equation. We wish to exercise **lap3d** for determining the resistance R using 3D FD. The reader needs the background of previous chapters. Let us go through the following example.

Problem 1:

It is given that the electromagnetic system of figure 4.5(a) has the gold plate electrode with $\sigma = 4.1 \times 10^7$ mhos/meter which renders the resistance $R = 1.2195 \times 10^{-7}$ Ω subject to $a = 1cm$, $b = 5cm$, and $w = 4cm$. We intend to verify the resistance using 3D FD tactic.

Solution:

You may choose any V_0 say $V_0 = 100V$. The 3D domain is formed by $0 \leq x \leq 5cm$, $-1cm \leq y \leq 0$, and $0 \leq z \leq 4cm$. If we consider 21 points in each coordinate direction, the resolution Δx is $5cm/20$ and enter the domain and resolution information to like name variables e.g. a to **a**:

```
>>a=1; b=5; w=4; P=20; dx=b/P; dy=a/P; dz=w/P; ↵
```

The user-chosen **P** in above holds the number of samples minus one. The $100V$ is in $y = -1cm$ plane in index domain which becomes $[m,1,k]$. Consider chapter 3 quoted maximum iteration number and mean square error and call the Laplace equation solver as follows:

```
>>f=lap3d([0 b -a 0 0 w],dx,dy,dz,'f(:,1,:)=100;',1e-10,25); ↵
```

Hence $f(x,y,z)$ samples are available in **f** as a 3D array. The negative gradient or electric field is given by $\overline{E} = -\nabla f(x,y,z)$ from which only E_y component gives some integration result and get it as a 3D array by (chapter 2):

```
>>Ey=-diff(f,1,1)/dy; ↵
```

The **Ey** is a user-chosen variable that holds E_y. The denominator $\oint \overline{E} \circ d\overline{S}$ becomes $\Sigma\Sigma\ E_y \Delta x \Delta z$ in FD domain hence compute the resistance by:

```
>>R=100/sum(sum(Ey(:,1,:)))/dx/dz/4.1e7/1e-2 ↵
```

```
R =
        1.1614e-007
```

Although E_y is a three dimensional array, we do not need all data just $E_y(x,-1cm,z)$ for the plate is involved in the computing which is **Ey(:,1,:)** and a rectangular matrix. Appendix B.9 mentioned **sum** is exercised two times for the double summation. The reader needs to pay attention to the units. The E_y is in V/cm and the term $\Sigma\Sigma\ E_y \Delta x \Delta z$ has unit $V.cm$ which has to be $V.m$ for this reason 10^{-2} or **1e-2** is used.

Some discrepancy is observed in 3D FD computing; $1.1614 \times 10^{-7}\ \Omega$ instead of $1.2195 \times 10^{-7}\ \Omega$. You may improve the result by taking more than 21 points say 31, 41, etc or reducing mean square error. Another important point is standard result has perfect long plate assumption which is not. It is not that FD achieved results are wrong instead very close to realistic structure based solution. Do not forget that higher sample numbers require longer time to execute so be patient during the run time.

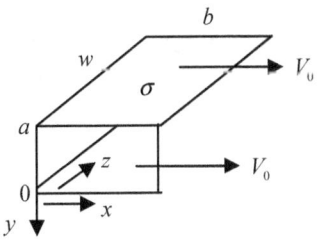

Figure 4.5(b) A 3D electromagnetic system with two constant boundary voltages

Problem 2:

In problem 1 now two plates form a L shaped electrode which are located on the planes $y = -1cm$ and $z = 0cm$ as in figure 4.5(b). Compute the resistance of the electromagnetic system.

Solution:

Choose $V_0 = 100V$ on $y = -1cm$ and $z = 0cm$ planes. The two boundary conditions are $f(x,-1cm,z) = 100V$ and $f(x,y,0cm) = 100V$ or $f[m,1,k] = 100V$ and $f[m,n,1] = 100V$ respectively. Enter the relevant parameters and get the $f(x,y,z)$ solution like problem 1 by:

```
>>a=1; b=5; w=4; P=20; dx=b/P; dy=a/P; dz=w/P; ↵
>>f=lap3d([0 b -a 0 0 w],dx,dy,dz,'f(:,1,:)=100;f(:,:,1)=100;',1e-10,25); ↵
```

This time we have two components of electric field $\overline{E} = -\nabla f(x,y,z)$ namely E_y and E_z, each of which contributes to the resistance computing therefore compute them by (section 2.8):

```
>>Ey=-diff(f,1,1)/dy; Ez=-diff(f,1,3)/dz; ↵
```

The user-chosen variables Ey and Ez retain the samples of E_y and E_z respectively, each as a 3D array.

The $R = \dfrac{V_0}{\oint \sigma \overline{E} \circ dS}$ approximates to $R = \dfrac{V_0}{\sigma \sum\limits_x \sum\limits_z E_y \, \Delta x \, \Delta z + \sigma \sum\limits_x \sum\limits_y E_z \, \Delta x \, \Delta y}$

for the given L plate electromagnetic system. The $E_y(x,-1cm,z)$ and $E_z(x,y,0cm)$ are understood to be in the denominator of R and pick them from Ey and Ez by Ey(:,1,:) and Ez(:,:,1) respectively. In order to avoid mistakes in coding, let us compute the denominator in two parts:

```
>>Ry=4.1e7*sum(sum(Ey(:,1,:)))*dx*dz*1e-2; ↵
>>Rz=4.1e7*sum(sum(Ez(:,:,1)))*dx*dy*1e-2; ↵
>>R=100/(Ry+Rz) ↵
```

R =

8.0646e-008

Therefore the resistance of the electromagnetic system in figure 4.5(b) is $R = 8.0646 \times 10^{-8} \Omega$. In last executions the user-supplied Ry and Rz hold the computed $\sigma \sum\limits_x \sum\limits_z E_y \, \Delta x \, \Delta z$ and $\sigma \sum\limits_x \sum\limits_y E_z \, \Delta x \, \Delta y$ in standard unit respectively. What do we infer from the last computing? As the plate area encompassing the electromagnetic system increases, resistance of the system decreases.

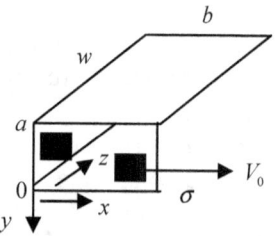

Figure 4.5(c) A 3D electromagnetic system with two segment electrodes

Problem 3:
Instead of a solid electrode two tiny electrode plates are attached on the front face of the problem 1 which is depicted in figure 4.5(c). The two plates are defined by $0.5cm \leq x \leq 1.4cm$, $-0.9cm \leq y \leq -0.5cm$ and $3.6cm \leq x \leq 4.5cm$, $-0.5cm \leq y \leq -0.1cm$ on the plane $z = 0cm$. Compute resistance of the electromagnetic system subject to problem 1 quoted parameters.

Solution:
In problem 1 we chose 21 point resolution for every coordinate. That does not suit here because $1.4\,cm$ bound of x is not multiple of $0.5\,cm$ as in the case in problem 1. A resolution of $0.1\,cm$ is suitable for every coordinate for which we need the sample point number as 51 along x whereas 11 and 21 along y and z respectively. Generate coordinate directed resolutions by:
>>a=1; b=5; w=4; dx=b/50; dy=a/10; dz=w/20; ⏎
The intervals $0.5cm \leq x \leq 1.4cm$, $-0.9cm \leq y \leq -0.5cm$, $3.6cm \leq x \leq 4.5cm$, and $-0.5cm \leq y \leq -0.1cm$ translate to [6,15], [2,6], [37,46], and [6,10] in FD domain respectively. Knowing so, call the solver then (section 3.5):
>>f=lap3d([0 b -a 0 0 w],dx,dy,dz,'f(6:15,2:6,1)=100;f(37:46,6:10,1)=100;',
    1e-10,25); ⏎
Compute the E_z by:
>>Ez=-diff(f,1,3)/dz; ⏎
For two segments of the electromagnetic system the resistance is computed by $R = \dfrac{V_0}{\sigma \sum\limits_x \sum\limits_y E_{z1} \Delta x \Delta y + \sigma \sum\limits_x \sum\limits_y E_{z2} \Delta x \Delta y}$ where E_{z1} and E_{z2} are electric fields in the first and second segments respectively so carry out the following:
>>D1=4.1e7*sum(sum(Ez(2:6,6:15,1)))*dx*dy*1e-2; ⏎
>>D2=4.1e7*sum(sum(Ez(6:10,37:46,1)))*dx*dy*1e-2; ⏎
In above the user-chosen D1 and D2 hold the contribution from E_{z1} and E_{z2} respectively from which the total resistance is computed by:
>>R=100/(D1+D2) ⏎

R =
        7.5779e-007
i.e. $R = 7.5779 \times 10^{-7}\,\Omega$ note that x and y coordinates are interchanged due to convention difference between 2D and 3D systems (chapter 2).

4.9 Capacitance computing using 3D FD
Capacitance computing is similar to that of the resistance but employs the expression $C = \dfrac{\oint \varepsilon \overline{E} \circ d\overline{S}}{V_0}$ where the quantities involved have the section 4.8 quoted meanings with the exception ε which is the permittivity of the dielectric between electrodes. The absolute permittivity ε is connected

to relative permittivity ε_r by $\varepsilon = \varepsilon_r \varepsilon_0$ where ε_0 is the free space permittivity $8.854\times10^{-12}\, Farad/meter$. For every problem in section 4.8 we are going to compute the capacitance with plastic dielectric $\varepsilon_r = 2.3$. The geometry involving permittivity is shown in figure 4.6(a) and now there is no need for conductivity σ.

Problem 1:

Figure 4.6(a) depicted electro-magnetic system is filled with plastic dielectric and geometric dimension of the system is taken from problem 1 of last section. Exercise 3D FD to verify the capacitance of the electromagnetic system which is given as $C = 4.0728\times10^{-12}\, F/m$.

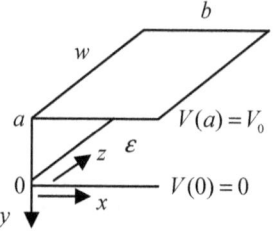

Figure 4.6(a) Geometry for capacitance of electromagnetic system in figure 4.5(a)

Solution:

Whatever we conducted in problem 1 until Ey of last section must be reexecuted:

```
>>a=1; b=5; w=4; P=20; dx=b/P; dy=a/P; dz=w/P; ↵
>>f=lap3d([0 b -a 0 0 w],dx,dy,dz,'f(:,1,:)=100;',1e-10,25); ↵
>>Ey=-diff(f,1,1)/dy; s=2.3*8.854e-12; ↵
```

In above user-chosen s holds $\varepsilon = 2.3\times8.854\times10^{-12}\, Farad/meter$. The capacitance expression becomes $C = \dfrac{\varepsilon \Sigma\Sigma\, E_y\, \Delta x\, \Delta z}{V_0}$ in the 3D FD domain and its straightforward execution is the following:

```
>>C=s*sum(sum(Ey(:,1,:)))*dx*dz*0.01/100 ↵

C =
     4.2765e-012
```

i.e. the capacitance is $C = 4.2765\times10^{-12}\, F/m$. The 0.01 multiplication is basically turning the electric field from V/cm to V/m. The discrepancy is because of FD computing nature like the resistance counterpart. Manipulating total sample number, maximum iteration number, or maximum mean square error you may get the result close to the exact one.

Problem 2:

The L shaped electrode (problem 2) of section 4.8 is now filled with plastic dielectric, compute capacitance of the electromagnetic system.

Solution:

For continuity we repeat the previous commands:

```
>>a=1; b=5; w=4; P=20; dx=b/P; dy=a/P; dz=w/P; ↵
>>f=lap3d([0 b -a 0 0 w],dx,dy,dz,'f(:,1,:)=100;f(:,:,1)=100;',1e-10,25); ↵
```

```
>>Ey=-diff(f,1,1)/dy; Ez=-diff(f,1,3)/dz; ↵
```

The capacitance is computed by $\dfrac{\varepsilon\sum\limits_{x}\sum\limits_{z} E_y\,\Delta x\,\Delta z + \varepsilon\sum\limits_{x}\sum\limits_{y} E_z\,\Delta x\,\Delta y}{V_0}$ so analogous

commands are the following:

```
>>s=2.3*8.854e-12; Cy=s*sum(sum(Ey(:,1,:)))*dx*dz*1e-2; ↵
>>Cz=s*sum(sum(Ez(:,:,1)))*dx*dy*1e-2; ↵
>>C=(Cy+Cz)/100 ↵
```

```
C =
        6.1589e-012
```

In last executions user-supplied variables **Cy** and **Cz** implement $\varepsilon\sum\limits_{x}\sum\limits_{z} E_y\,\Delta x\,\Delta z$ and $\varepsilon\sum\limits_{x}\sum\limits_{y} E_z\,\Delta x\,\Delta y$ respectively. The factor **1e-2** or 0.01 is for the conversion of V/cm to V/m therefore capacitance of the electromagnetic system in figure 4.5(b) is $C = 6.1589\times10^{-12} F$.

Problem 3:

Two tiny electrode plates of figure 4.5(c) are now filled with plastic dielectric, compute capacitance of the electromagnetic system.

Solution:

Most commands of last section explained problem 3 are needed as follows:

```
>>a=1; b=5; w=4; dx=b/50; dy=a/10; dz=w/20; s=2.3*8.854e-12; ↵
>>f=lap3d([0 b -a 0 0 w],dx,dy,dz,'f(6:15,2:6,1)=100;f(37:46,6:10,1)=100;',
        1e-10,25); ↵
>>Ez=-diff(f,1,3)/dz; ↵
```

For the two segments capacitance is computed by $C = \dfrac{\varepsilon\sum\limits_{x}\sum\limits_{y} E_{z1}\,\Delta x\,\Delta y}{V_0} +$

$\dfrac{\varepsilon\sum\limits_{x}\sum\limits_{y} E_{z2}\,\Delta x\,\Delta y}{V_0}$ and its relevant execution is carried out in the following:

```
>>C1=s*sum(sum(Ez(2:6,6:15,1)))*dx*dy*1e-2; ↵
>>C2=s*sum(sum(Ez(6:10,37:46,1)))*dx*dy*1e-2; ↵
```

In above the user-chosen **C1** and **C2** hold the contributions from E_{z1} and E_{z2} respectively from which the total capacitance is computed by:

```
>>C=(C1+C2)/100 ↵
```

```
C =
        6.5544e-013
```

i.e. $C = 6.5544\times10^{-13}\,Farad$ which is indicative of reduction due to increased plate area.

4.10 3D FD on composite structures

So far we have been addressing the 3D space containing single electrode or dielectric in a Cartesian electromagnetic system. What if we have a system which comprises of two different electrodes or dielectrics? Treatment required for this sort of problems is addressed in this section.

Problem 1:

Figure 4.5(c) depicted two tiny electrodes are composed of gold and aluminium plates which have conductivities $\sigma_1 = 4.1 \times 10^7$ mhos/meter and $\sigma_2 = 3.5 \times 10^7$ mhos/meter respectively. Compute resistance of the two dissimilar plate electromagnetic system using 3D FD.

Solution:

Repeat commands of the problem 3 in section 4.8:
```
>>a=1; b=5; w=4; dx=b/50; dy=a/10; dz=w/20; ↵
>>f=lap3d([0 b -a 0 0 w],dx,dy,dz,'f(6:15,2:6,1)=100;f(37:46,6:10,1)=100;',
    1e-10,25); ↵
>>Ez=-diff(f,1,3)/dz; ↵
```

The $\dfrac{V_0}{\sigma_1 \sum\limits_{x} \sum\limits_{y} E_{z1} \Delta x \Delta y + \sigma_2 \sum\limits_{x} \sum\limits_{y} E_{z2} \Delta x \Delta y}$ is used to compute the resistance as far

as the plate circumstance is concerned, its implementation is as follows:
```
>>s1=4.7e7; s2=3.5e7; ↵
>>D1=s1*sum(sum(Ez(2:6,6:15,1)))*dx*dy*1e-2; ↵
>>D2=s2*sum(sum(Ez(6:10,37:46,1)))*dx*dy*1e-2; ↵
>>R=100/(D1+D2) ↵
```

```
R =
        7.5779e-007
```

i.e. $R = 7.5779 \times 10^{-7} \Omega$, new workspace variables s1 and s2 are user-chosen and retain σ_1 and σ_2 respectively.

Problem 2:

Figure 4.6(b) depicted dielectric is divided equally in two portions with relative permittivities $\varepsilon_{r1} = 2.3$ and $\varepsilon_{r1} = 3$. All other specifications are taken from section 4.8. Compute capacitance of the electromagnetic system using 3D FD.

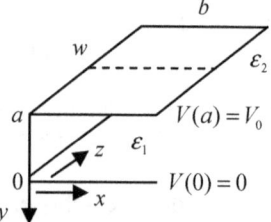

Figure 4.6(b) Geometry for capacitance of a dual dielectric system

Solution:

First reexecute the earlier commands until you get Ey then enter the absolute permittivities ε_1 and ε_2 to user-chosen e1 and e2 respectively as follows:

```
>>a=1; b=5; w=4; P=20; dx=b/P; dy=a/P; dz=w/P; ↵
>>f=lap3d([0 b -a 0 0 w],dx,dy,dz,'f(:,1,:)=100;',1e-10,25); ↵
>>Ey=-diff(f,1,1)/dy; e1=2.3*8.854e-12; e2=3*8.854e-12; ↵
```

The capacitance expression we employ is $C = \dfrac{\varepsilon_1 \sum\sum E_{y1} \Delta x \Delta z + \varepsilon_2 \sum\sum E_{y2} \Delta x \Delta z}{V_0}$

in the 3D FD domain and its straightforward execution is the following:

```
>>Cy1=e1*sum(sum(Ey(:,1,1:11)))*dx*dz*0.01/100; ↵
>>Cy2=e2*sum(sum(Ey(:,1,12:21)))*dx*dz*0.01/100; ↵
>>C=Cy1+Cy2 ↵
```

```
C =
     4.8963e-012
```

In above user-chosen variables **Cy1** and **Cy2** implement $\dfrac{\varepsilon_1 \sum\sum E_{y1} \Delta x \Delta z}{V_0}$ and

$\dfrac{\varepsilon_2 \sum\sum E_{y2} \Delta x \Delta z}{V_0}$ respectively from which total capacitance **C** is calculated by

the summation of **Cy1** and **Cy2** i.e. $C = 4.8963 \times 10^{-12} F/m$. There are 21 planes in z direction, we assume that approximately planes at $z=1$ through 11 form first half of the plate and the ones at $z=12$ through 21 do the rest half.

4.11 Divergence and curl of electric field in 3D space

In 3D space the divergence $D(x,y,z)$ of electric field $\overline{E}(x,y,z) = E_x(x,y,z)\overline{a}_x + E_y(x,y,z)\overline{a}_y + E_z(x,y,z)\overline{a}_z$ in Cartesian system is defined as $D = \nabla \bullet \overline{E} = \dfrac{\partial E_x}{\partial x} + \dfrac{\partial E_y}{\partial y} + \dfrac{\partial E_z}{\partial z}$. MATLAB embedded function **divergence** implements the computing which requires the syntax **divergence**(x-only grid points as a 3D array, y-only grid points as a 3D array, z-only grid points as a 3D array, $E_x(x,y,z)$ as a 3D array, $E_y(x,y,z)$ as a 3D array, $E_z(x,y,z)$ as a 3D array) and the return from **divergence** is $D(x,y,z)$ as a 3D array.

Again in 3D space the curl $\overline{C}(x,y,z)$ of electric field $\overline{E}(x,y,z) = E_x(x,y,z)\overline{a}_x + E_y(x,y,z)\overline{a}_y + E_z(x,y,z)\overline{a}_z$ in Cartesian system is defined as $\overline{C} =$

$\nabla \times \overline{E} = \begin{vmatrix} \overline{a}_x & \overline{a}_y & \overline{a}_z \\ \dfrac{\partial}{\partial x} & \dfrac{\partial}{\partial y} & \dfrac{\partial}{\partial z} \\ E_x & E_y & E_z \end{vmatrix}$. MATLAB embedded function **curl** implements

the computing which requires the syntax $[C_x(x,y,z), C_y(x,y,z), C_z(x,y,z)] =$ **curl**(x-only grid points as a 3D array, y-only grid points as a 3D array, z-only grid points as a 3D array, $E_x(x,y,z)$ as a 3D array, $E_y(x,y,z)$ as a 3D array, $E_z(x,y,z)$ as a 3D array) and the return from **curl** is $\overline{C}(x,y,z)$ as a 3D array. The $C_x(x,y,z)$, $C_y(x,y,z)$, and $C_z(x,y,z)$ are the three vector components of $\overline{C}(x,y,z)$ each as a 3D array.

All elements in divergence or curl computing must be identical size 3D array.

Divergence computing:

In section 4.4 we mentioned an example for computing electric field $\overline{E}(x,y,z)$ whose three components are stored in **Ex**, **Ey**, and **Ez**. Let us compute the divergence $D = \nabla \bullet \overline{E}$ of that electric field using 3D FD.

Reexecute the previous commands until you obtain identical size **Ex**, **Ey**, and **Ez** and generate the basic grid points by:

```
>>x=0:dx:10; y=0:dy:6; z=-5:dz:0; [X,Y,Z]=meshgrid(x,y,z); ↵
```

Invoke the embedded function then:

```
>>D=divergence(X,Y,Z,Ex,Ey,Ez); ↵
```

In above user-chosen **D** holds $D(x,y,z)$ samples as a 3D array.

Curl computing:

Compute curl $\overline{C} = \nabla \times \overline{E}$ by 3D FD for just mentioned electric field.

All computing elements we need are available from the divergence computing, just call the embedded function:

```
>>[Cx Cy Cz]=curl(X,Y,Z,Ex,Ey,Ez); ↵
```

In above user-chosen **Cx**, **Cy**, and **Cz** hold $C_x(x,y,z)$, $C_y(x,y,z)$, and $C_z(x,y,z)$ each as a 3D array respectively.

Divergence graphing:

The $D(x,y,z)$ is a scalar function like potential function $f(x,y,z)$ of section 4.2. Whatever graph we implemented in the section can be exercised on above **D** or $D(x,y,z)$.

Curl graphing:

The $\overline{C}$ is a vector function like electric field function $\overline{E}(x,y,z)$ of section 4.4. Whatever graph we implemented in the section can be exercised on above **[Cx Cy Cz]** or $\overline{C}$.

4.12 Four dimensional visualizations of Laplace equation solution

We could have addressed this section earlier in this chapter but complexity involvement prevents us from doing so. This is basically case 3 of section 4.1. What is the complexity here? We have now volume type data. What is inside the volume is not visible from outside. Depending on the viewing angle there are infinite views and every view is perceived differently. Perfectly speaking we need 360^0 vision system which computer monitor is not. Yet there are some tools which may help us view the solution or its related quantities from some context of 4D. Importantly we have data for x, y, z, and $f(x,y,z)$ each as a 3D array - that is why it is called 4D.

Problem 1: All image plots for potential in the whole 3D space

Consider the example 1 of section 4.2. The potential solution for $f(x,y,z)$ is available in variable **f**. We wish to view $f(x,y,z)$ solution as image plot for all z.

Solution:

From chapter 2 we know that $f(x,y,z)$ turns to $f[m,n,k]$. In essence we intend to view $f[m,n,1]$, $f[m,n,2]$, $f[m,n,3]$, etc as digital image but for all k. In section 4.3 we explained how to display $f[m,n,k]$ as an image for a particular k. There are 21 indexes for k i.e. the third dimension of **f** in MATLAB workspace browser.

The way we solve the problem is form a four dimensional array **D** (some user-chosen variable), the first three dimensions of which refer to $f[m,n,k]$ for every k. By adding some colormap artificially we turn the 2D type $f[m,n,k]$ with frozen k to 3D and assign that to first three dimensions of **D**. The fourth dimension of **D** holds the particular k. Anyhow the whole programming circumstance is as follows:

```
for k=1:21
    D(:,:,:,k)=ind2rgb(im2uint8(mat2gray(f(:,:,k))),pink);
end
montage(D)
```

The f(:,:,k) means the k-th plane in the z direction. The embedded function **mat2gray** maps any data between 0 and 1. The **im2uint8** transforms 0-1 data into integer [0,255] for image display reason. The **pink** is an artificial colormap taken from MATLAB. The embedded function **ind2rgb** transforms the 2D data in conjunction with artificial colormap into a 3D array for JPEG (a popular digital image format) or RGB (short for red-green-blue) image format. The whole 3D element is assigned to the k-th element of 4D array **D** i.e. **D(:,:,:,k)**. The **montage** is an embedded function which displays all elements of a 4D array as an image in row directed way like figure 4.7(a). Left uppermost corner of the figure refers to

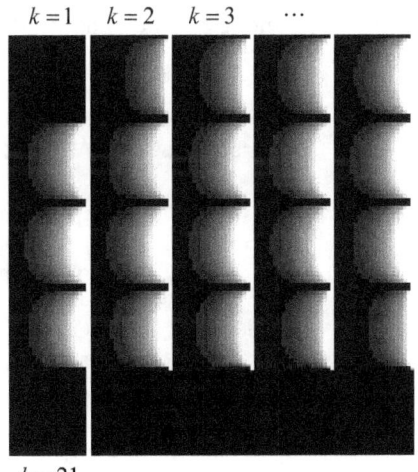

$k=1$ $k=2$ $k=3$ $\cdots$

$k=21$

Figure 4.7(a) Image plot of $f[m,n,k]$ for different k

$k=1$. There are total 5×5 image elements in figure 4.7(a). The first row from the top shows images for $k = 1$ through 5 thus $k=21$ is displayed in the fifth row. The images indexed by $k=22$ through 25 are left empty or full black (last row in the figure).

Sorry folks the text is written in black and white form, MATLAB displays graphics in color form so do not get disappointed. Regarding scale in every image the white zone is for maximum, black is for minimum, and gray is between maximum and minimum on the potential values of $f(x,y,z)$.

Indeed this is not a 4D visualization but one point is sure that we can have an overview on the whole solution space - the way $f(x,y,z)$ concentrates in the solution space.

Problem 2: All image plots for electric field in the whole 3D space

What about the electric field solution? Certainly you can and the electric field is a vector quantity and visualize it component by component. Usually magnitude of an electric field is visualized. Consider section 4.4 quoted electric field $\bar{E}(x,y,z)=E_x(x,y,z)\bar{a}_x+E_y(x,y,z)\bar{a}_y+E_z(x,y,z)\bar{a}_z$ whose components are stored in workspace variables Ex, Ey, and Ez. We wish to view $|\bar{E}(x,y,z)|$ solution as an image plot for all z like problem 1.

Solution:

Repeat section 4.4 cited commands until you obtain identical size Ex, Ey, and Ez. The magnitude electric field $|\bar{E}(x,y,z)|$ is obtained from $\sqrt{E_x^2(x,y,z)+E_y^2(x,y,z)+E_z^2(x,y,z)}$, computing of which requires the scalar code clearly as a 3D array (appendix A):

```
>>E=sqrt(Ex.^2+Ey.^2+Ez.^2); ↵
```

Then repeat the problem 1 illustrated commands with E i.e.

```
for k=1:21
        D(:,:,:,k)=ind2rgb(im2uint8(mat2gray(E(:,:,k))),pink);
end
montage(D)
```

The graph is not shown for space reason but similar to figure 4.7(a). In every image you will find the gray zone expanded owing to the nature of electric field.

Problem 3: All 2D plots for potential/field in the whole 3D space

It is possible to view one directional variation of electric potential or field by dint of the embedded function subplot over the whole solution space (appendix C). Consider the problem 1 of section 4.2. We graphed $f(x,3cm,-2cm)$ versus x in the section. Now we wish to view $f(x,3cm,z)$ versus x for all z in the solution space of $f(x,y,z)$.

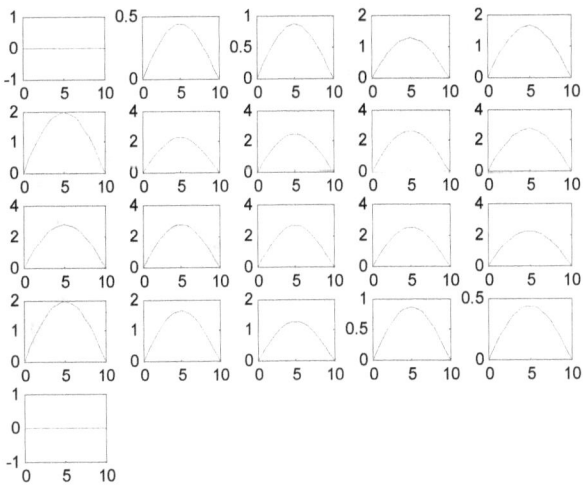

Figure 4.7(b) $f(x,3cm,z)$ versus x for all z

Solution:

Repeat commands of the section for solution of $f(x,y,z)$:

```
>>dx=10/20; dy=6/20; dz=5/20; ⏎
>>f=lap3d([0 10 0 6 -5 0],dx,dy,dz,'f(21,:,:)=100;',1e-10,25); x=0:dx:10; ⏎
```

Now exercise the **subplot** and **plot** for $f(x,3cm,z)$ versus x in conjunction with a for-loop (appendix B.4) as follows:

```
>>for k=1:21, subplot(5,5,k), plot(x,f(:,11,k)), end ⏎
```

The $f(x,3cm,z)$ becomes $f[m,11,k]$ in index domain and the for-loop counter k controls every single k. Outcome of the last line command is shown in figure 4.7(b). We have chosen 5×5 subwindow for the **subplot** which is why the input argument of **subplot** is so. The correspondence of k is similar to the one in figure 4.7(a).

This is not a 4D plot but by changing the second index of f e.g. f(:,12,k) you can graph for the next y plane thus covering the whole 3D solution space.

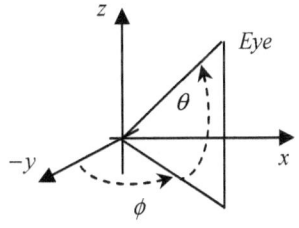

Figure 4.7(c) Geometry of the view function

Problem 4: Viewing surface graph at different angles in 3D space

Upon graphing a surface plot, viewing angle might be changed to see potential or field solution from another perspective. This has close link with realistic 4D graphing. Make no mistake there are infinite viewing angles. Anyhow figure 4.7(c) shows geometry of the eye viewing angle in MATLAB graphics which is (ϕ,θ). First draw the surface graph using section 4.3 cited functional tools then exercise the embedded function view with the syntax view(ϕ,θ) where each angle is in degree and user-chosen. Figure 4.2(c) depicts a surface graph, certainly that is displayed based on a default viewing angle. We wish to perceive the graph when (ϕ,θ) =($25^0,57^0$).

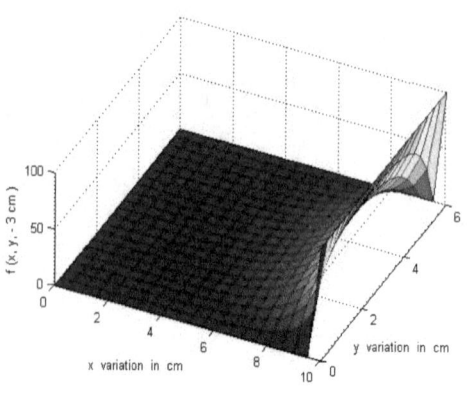

Figure 4.7(d) Surface graph of figure 4.2(c) viewed at (ϕ,θ)=($25^0,57^0$)

Solution:

Repeat commands of sections 4.2 and 4.3 for figure 4.2(c):
```
>>dx=10/20; dy=6/20; dz=5/20; ↵
>>f=lap3d([0 10 0 6 -5 0],dx,dy,dz,'f(21,:,:)=100;',1e-10,25); ↵
>>x=0:dx:10; y=0:dy:6; ↵
>>fxy=f(:,:,13); ↵
>>surf(x,y,fxy) ↵
```
Then change the default view angle by calling:
```
>>view(25,57) ↵
```
The response is shown in figure 4.7(d), definitely the perception is different. Different (ϕ,θ) results different perceptions which you can check easily.

Problem 5: Potential/field distribution along planes in 3D space

We may insert a plane inside the solution space and view the distribution of potential or field across the plane. Embedded function slice is useful in this regard. Concerning the section 4.2, consider the solution of potential $f(x,y,z)$ which is stored in workspace f. We intend to view the potential $f(x,y,z)$ distribution at planes $z=-2.5cm$ and $z=-4cm$.

Solution:

The syntax we employ is slice(x -only grid points as a 3D array, y - only grid points as a 3D array, z -only grid points as a 3D array, $f(x,y,z)$ as a 3D array, wanted x planes as a row matrix, wanted y planes as a row matrix, wanted z planes as a row matrix). Absence of any plane is fed as empty or [ ]. Repeat the section 4.2 quoted commands:
```
>>dx=10/20; dy=6/20; dz=5/20; ↵
>>f=lap3d([0 10 0 6 -5 0],dx,dy,dz,'f(21,:,:)=100;',1e-10,25); ↵
```
Generate the directed and merely grid points by:

```
>>x=0:dx:10; y=0:dy:6; z=-5:dz:0; [X,Y,Z]=meshgrid(x,y,z); ↵
```
Invoke the distribution plotter by:
```
>>slice(X,Y,Z,f,[ ],[ ],[-2.5 -4]) ↵
>>colorbar ↵
```

There is no mention about x or y planes that is why the 5th and 6th input arguments of **slice** are empty. The 7th input argument is just the wanted z plane locations as a row matrix.

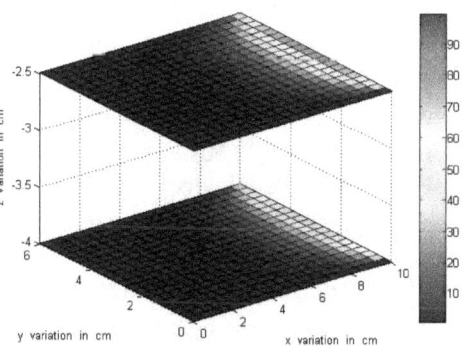

Figure 4.7(e) depicts the two potential distributions at the two planes. The command **colorbar** attaches a color code bar in the last drawn graph which indicates potential distribution values across the two planes. Sorry folks, the text

Figure 4.7(e) Potential distribution at two planes

is written in black and white form that is why every color image appears in gray form. You will find graphics as well as bar in color in MATLAB.

Problem 6: Electric field lines along different planes in 3D space

Consider section 4.4 quoted electric field $\overline{E}(x,y,z) = E_x(x,y,z)\overline{a}_x + E_y(x,y,z)\overline{a}_y + E_z(x,y,z)\overline{a}_z$ whose components are stored in workspace variables **Ex**, **Ey**, and **Ez**. We wish to view electric flux lines of $\overline{E}(x,y,z)$ at planes $y = 1cm$ and $y = 5cm$.

Solution:

In section 4.4 we explained the vector field based on a single plane cut. In this problem we bring 3D perspective on the electric field lines which call for multiple functional tools namely **box**, **daspect**, **view**, **axis**, **streamslice**, and **streamline**. Almost all these functional elements need grid points as a 3D array, let us carry out the following:
```
>>dx=10/20; dy=6/20; dz=5/20; ↵
>>x=0:dx:10; y=0:dy:6; z=-5:dz:0; ↵
```
That is we generated the directed grid points e.g. **x** for x. Each as a 3D array, the x-only, y-only, and z-only grid points are generated by (e.g. **X** for x):
```
>>[X,Y,Z]=meshgrid(x,y,z); ↵
```
Then execute the following:
```
>>grid; box; daspect([1 1 1]) ↵
```
The command **grid** opens a figure window and adds directed grid lines on that. The **box** adds a rectangular 3D box based on the graphics data. The **daspect** connects the user-decided aspect ratio to the figure window which

Mohammad Nuruzzaman

takes three element row matrix, every element of which is a directed scale factor i.e. daspect([S_x S_y S_z]). The S_x, S_y, and S_z are the user-chosen x, y, and z directed scale factors respectively. Every scale factor dictates contraction or expansion of the axis e.g. S_x along x. For axes setting according to given x, y, and z intervals, choose S_x=1, S_y=1, and S_z=1 which is what is conducted in last command line. See problem 4 for view and we chose (ϕ,θ)=($30^0,50^0$) that is:

>>view(30,50) ⌐

From the given intervals of x, y, and z, force the graphical axes to be according to (section 4.7 for axis):

>>axis([0 10 0 6 -5 0]) ⌐

The streamslice determines projection of $\overline{E}(x,y,z)$ on user-decided planes which requires the syntax streamslice(x-only grid point, y-only grid point, z-only grid point, $E_x(x,y,z)$ as a 3D array, $E_y(x,y,z)$ as a 3D array, $E_z(x,y,z)$

as a 3D array, wanted x planes as a row matrix, wanted y planes as a row matrix, wanted z planes as a row matrix). Absence of any plane is fed as empty or [ ]. The return from streamslice is a vector which has magnitude and phase or line and line direction in graphics term hence output argument of the

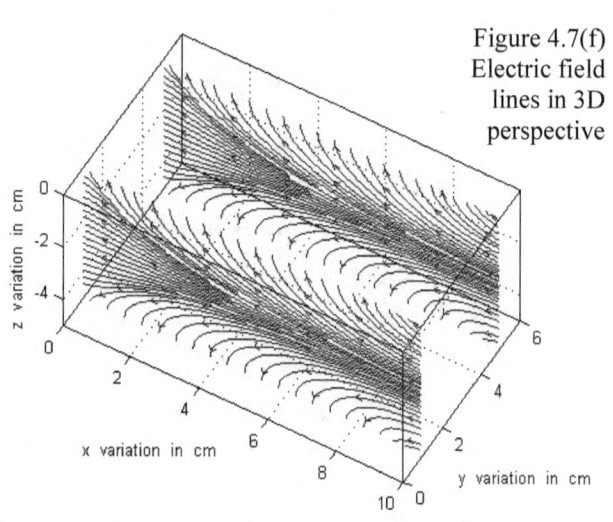

Figure 4.7(f) Electric field lines in 3D perspective

streamslice should be [M P] where M and P indicate magnitude and phase of line elements in the vector graph and user-chosen respectively. The vector information obtained in M and P is put to 3D perspective by the command streamline as follows:

>>[M P]=streamslice(X,Y,Z,Ex,Ey,Ez,[ ],[1 5],[ ]); ⌐
>>streamline([M P]) ⌐

The execution product you find in figure 4.7(f), see appendix C for peripheral properties e.g. x axis label.

Anyhow we terminate the chapter with this example.

Exercises

1. Figure 4.1(a) depicted rectangular box contains a potential $75\,V$ on the right side plane while the other sides are at $0V$. The box is formed by $0 \le x \le 6cm$, $0 \le y \le 9cm$, and $-3cm \le z \le 0$. Consider 31 points in each coordinate direction, determine the potential solution using 3D FD, and graph the following: (a) $f(x,6cm,-2cm)$ versus x and (b) $f(4cm,6cm,z)$ versus z. Assume the mean square error as 10^{-10} and maximum iteration number as 25 in the FD computing.

2. In problem 1 graph the scalar potential as a contour plot for the following: (a) $f(x,y,-1.5cm)$ versus x and y (b) $f(5cm,y,z)$ versus y and z.

3. Graph the scalar potential of problem 1 as a surface plot for the following: (a) $f(x,y,-1.5cm)$ versus x and y (b) $f(5cm,y,z)$ versus y and z.

4. Graph the scalar potential of problem 1 as a digital image plot for the following: (a) $f(x,y,-1.5cm)$ versus x and y (b) $f(5cm,y,z)$ versus y and z.

5. Plot the vector field lines of following electric fields for problem 1 mentioned scalar potential: (a) $E_x(x,y,-2cm)\overline{a}_x + E_y(x,y,-2cm)\overline{a}_y$ (b) $E_x(x,0cm,z)\overline{a}_x + E_z(x,0cm,z)\overline{a}_z$.

6. Plot the electric flux density lines $D_x(5cm,y,z)\overline{a}_x + D_y(5cm,y,z)\overline{a}_y$ for problem 1 mentioned scalar potential function.

7. Superimpose the electric potential lines of $f(5cm,y,z)$ on the flux line graph of problem 6.

8. Plot $d\overline{S}$ on the potential surface of problem 3(b).

9. The electromagnetic system of figure 4.5(a) has the metallic plate electrode with $\sigma = 3 \times 10^6$ mhos/meter. Compute the resistance using 3D FD subject to $a = 2cm$, $b = 3cm$, and $w = 3cm$. Consider 0.1 cm for each resolution.

10. The L shaped electrode of figure 4.5(b) has the specifications of problem 9. Compute resistance of the electromagnetic system using 3D FD.

11. The two tiny electrode of figure 4.5(c) has the specifications of problem 9 with electrode definition $0.2cm \le x \le 1cm$, $-1.8cm \le y \le -1cm$ and $2.2cm \le x \le 2.8cm$, $1cm \le y \le -0.2cm$ on $z = 0$ plane. Compute resistance of the electromagnetic system using 3D FD.

12. The electromagnetic system of figure 4.6(a) has the specifications of problem 9 and is filled with a dielectric of relative permittivity $\varepsilon_r = 5$. Compute capacitance of the electromagnetic system using 3D FD.

13. The L shaped electrode of figure 4.5(b) has the specifications of problem 9 and is filled with a dielectric of relative permittivity $\varepsilon_r = 5$. Compute capacitance of the electromagnetic system using 3D FD.

14. The two tiny electrode of problem 11 is filled with a dielectric of relative permittivity $\varepsilon_r = 5$. Compute capacitance of the electromagnetic system using 3D FD.

15. The dual dielectric electromagnetic system of figure 4.6(b) has the specifications of problem 9 and is filled with dielectrics of relative permittivities $\varepsilon_{r1} = 5$ and $\varepsilon_{r2} = 7$. Compute capacitance of the electromagnetic system using 3D FD.

16. Consider the problem 1 electric potential solution of $f(x, y, z)$, graph the potential distribution of $f(x, y, z)$ at three planes $y = 3\ cm$, $y = 5\ cm$, and $y = 7\ cm$.

17. Consider problem 1 quoted electric field $\overline{E}(x, y, z)$, compute the field using 3D FD, and display the vector electric field lines at planes $x = 1cm$, $x = 3cm$, and $x = 5cm$.

18. Consider the problem 1 electric potential solution of $f(x, y, z)$ and graph the $f(x, y, z)$ solution as equipotential line for all y.

Answers:

(1) (a) Figure A4.1   (b) Figure A4.2

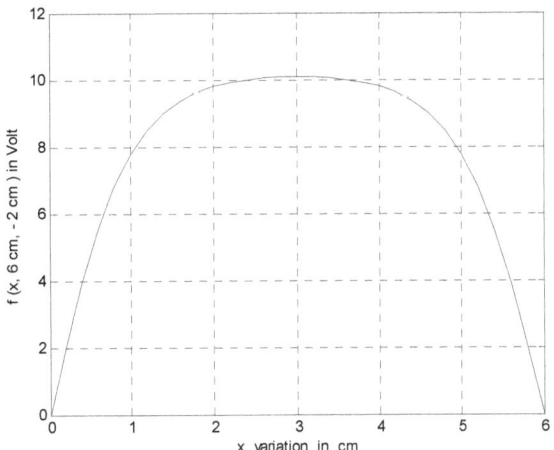

Figure A4.1 Plot of $f(x,6cm,-2cm)$ versus x

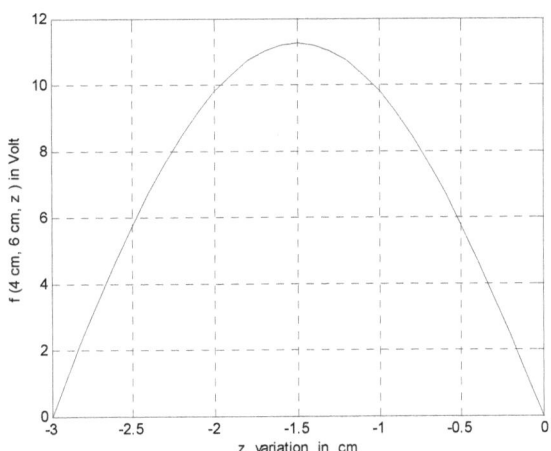

Figure A4.2 Plot of $f(4cm,6cm,z)$ versus z

Hint: section 4.2

(2) (a) Figure A4.3   (b) Figure A4.4

Hint: section 4.3

Figure A4.3 Contour plot of $f(x, y, -1.5cm)$ versus x and y - right side figure

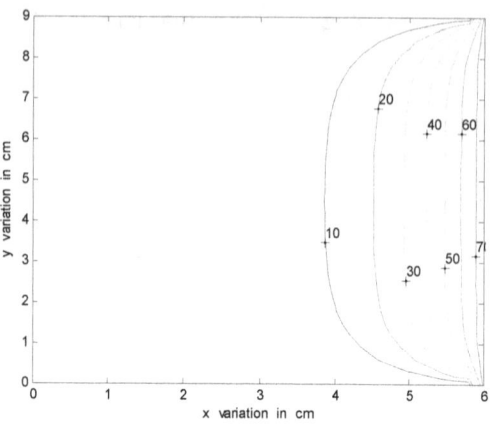

Figure A4.4 Contour plot of $f(5cm, y, z)$ versus y and z - right side figure

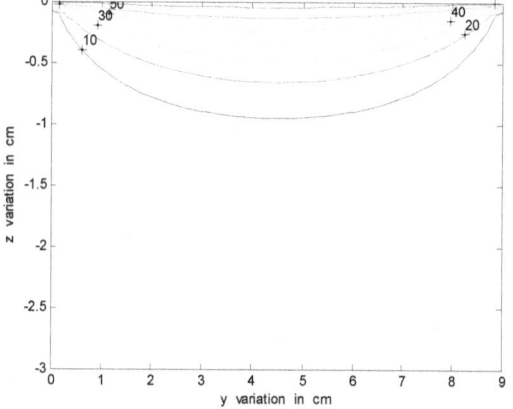

Figure A4.5 Surface plot of $f(x, y, -1.5cm)$ versus x and y - right side figure

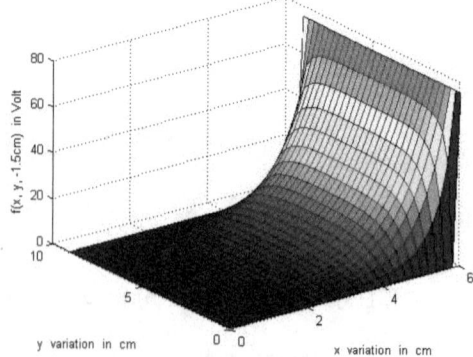

(3) (a) Figure A4.5   (b) Figure A4.6
     Hint: section 4.3

Figure A4.6 Surface plot of $f(5cm, y, z)$ versus y and z - right side figure

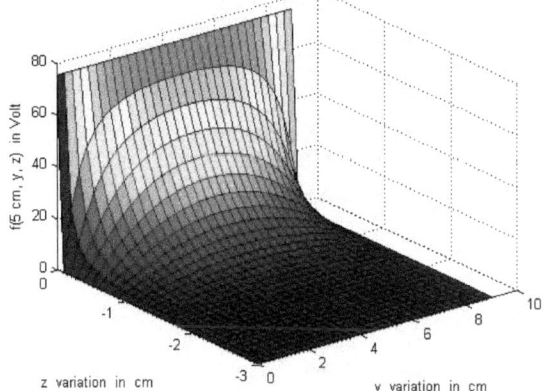

(4) (a) Similar to figure 4.2(d)
  (b) Figure A4.7
  Hint: section 4.3
(5) (a) Figure A4.8   (b) Figure A4.9
  Hint: section 4.4
(6) and (7) Figure A4.10
  Hint: sections 4.5 and 4.6
(8) Figure A4.11
  Hint: section 4.7
(9) $R = 1.0753 \times 10^{-5} \Omega$ considering the maximum iteration number and mean square error of section 4.8
  Hint: section 4.8
(10) $R = 1.7247 \times 10^{-6} \Omega$ considering the maximum iteration number and mean square error of section 4.8.
  Hint: section 4.8

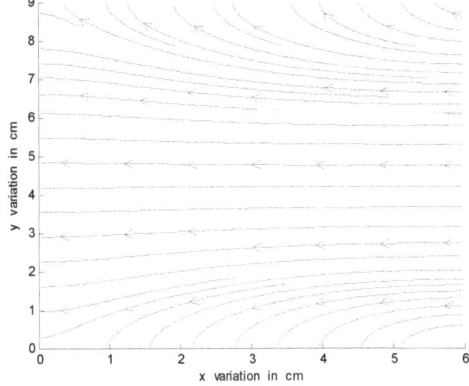

$\uparrow$
z

$y \rightarrow$

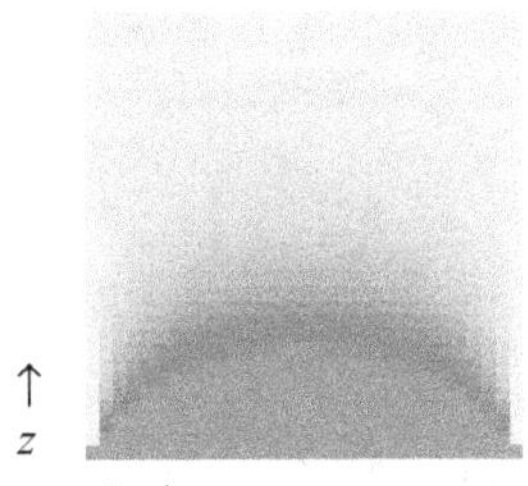

Figure A4.7 Image plot of $f(5cm, y, z)$ versus y and z

Figure A4.8 Vector electric field lines of $E_x(x, y, -2cm)\overline{a}_x + E_y(x, y, -2cm)\overline{a}_y$ versus x and y - right side figure

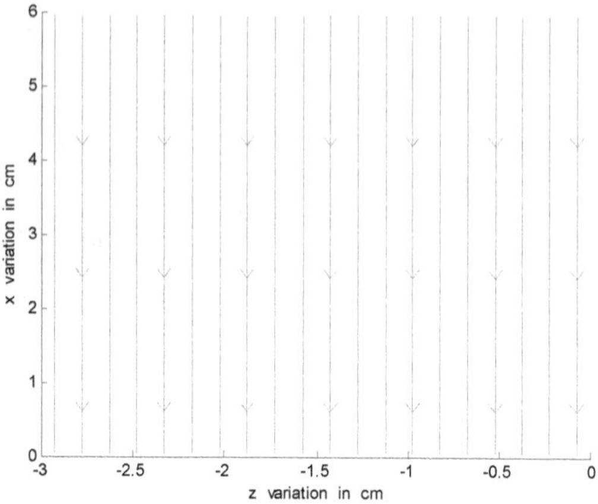

Figure A4.9 Vector
electric field lines of
$E_x(x, 0cm, z)\overline{a}_x +$
$E_z(x, 0cm, z)\overline{a}_z$ versus z
and x - right side figure

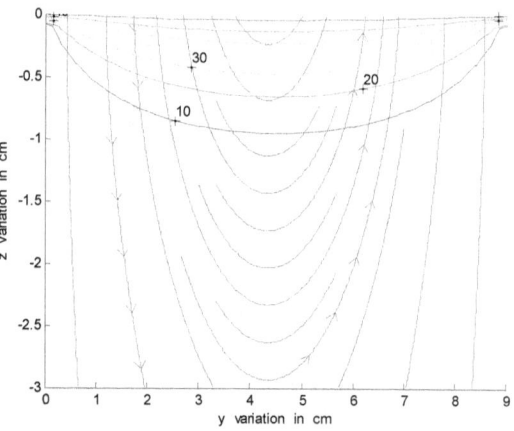

Figure A4.10 Electric
potential and field lines
together - right side
figure

Figure A4.11 Plot of
elementary surface
vector $d\overline{S}$ - right side
figure

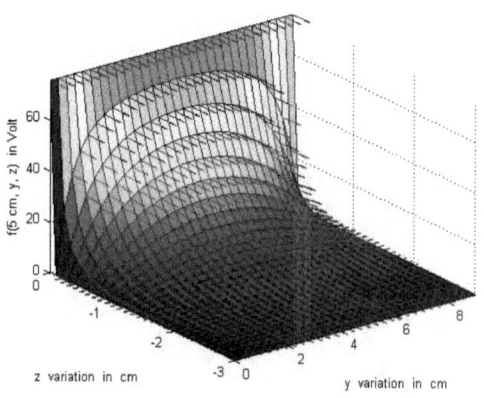

(11) $R = 6.4082 \times 10^{-6} \, \Omega$ considering the maximum iteration number and mean square error of section 4.8.   Hint: section 4.8

(12) $C = 1.3724 \times 10^{-12} \, F$ considering the maximum iteration number and mean square error of section 4.9.   Hint: section 4.9

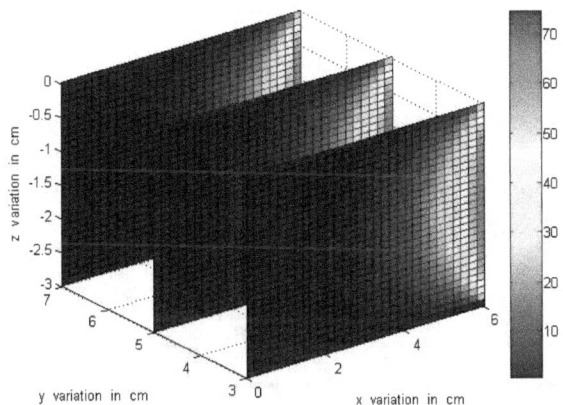

Figure A4.12 Potential distribution at three planes

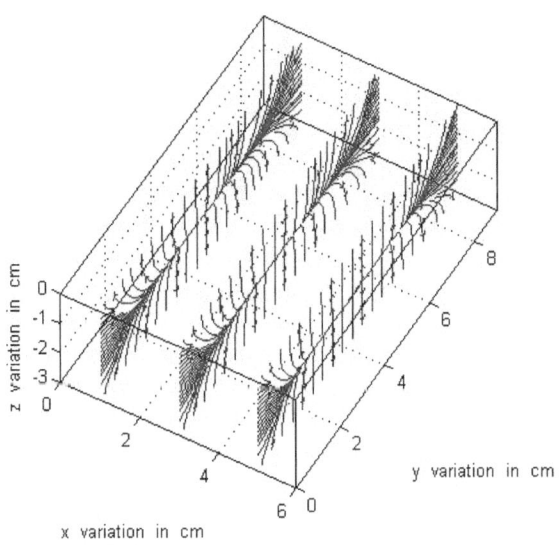

Figure A4.13 Electric field lines in 3D perspective at three planes

(13) $C = 8.5561 \times 10^{-12} F$ considering the maximum iteration number and mean square error of section 4.9.   Hint: section 4.9

(14) $C = 2.3028 \times 10^{-12} F$ considering the maximum iteration number and mean square error of section 4.9.   Hint: section 4.9

(15) $C = 1.6557 \times 10^{-12} F$ considering the maximum iteration number and mean square error of section 4.10.   Hint: section 4.10

(16) Figure A4.12   Hint: section 4.12

(17) Figure A4.13   Hint: section 4.12

(18) Figure A4.14   Hint: section 4.12

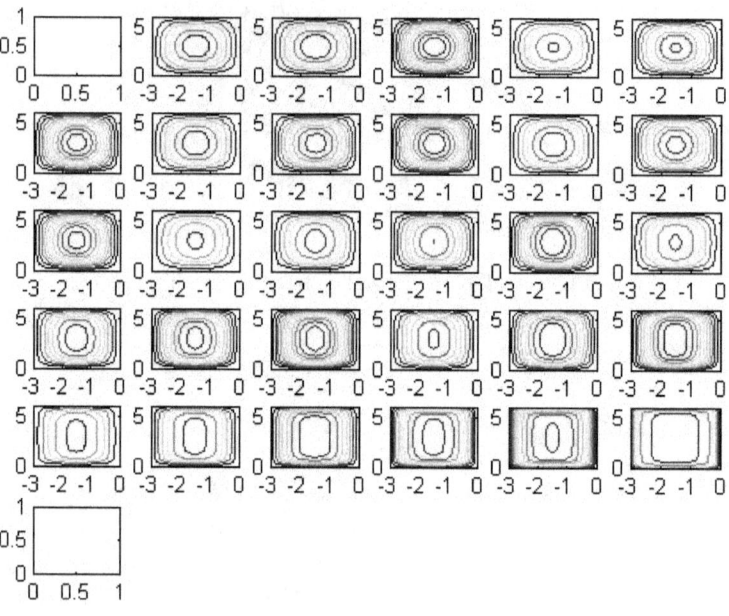

Figure A4.14 Equipotential line plot in
whole solution space

Appendices

Mohammad Nuruzzaman

Appendix A

Coding in MATLAB

MATLAB executes the code of an expression in terms of string which is the set of keyboard characters placed consecutively. One distinguishing feature of MATLAB is that the workspace variable itself is a matrix. The strings adopted for computation are divided into two classes - scalar and vector. The scalar computation results the order of the output matrix same as that of the variable matrix. On the contrary, the order for the vector computation is determined in accordance with the matrix algebra rules. Some symbolic functions and their MATLAB counterparts are presented in table A.1. The operators for arithmetic computations are as follows:

addition	+
subtraction	−
multiplication	*
division	/
power	^

The operation sequence of different operators in a scalar or vector string observes the following order:

enclosing braces	( )	first,
power operator	^	then,
division operator	/	next,
multiplication operator	*	after that,
addition operator	+	then, and
subtraction operator	−	finally.

The syntax of the scalar computation urges us to use .*, ./, and .^ in lieu of *, /, and ^ respectively. The operators *, /, and ^ are never preceded by . for the vector computation. The vector string is the MATLAB code of any symbolic expression or function often found in mathematics. In the sequel we present some examples on writing an expression in MATLAB.

◆ **Write MATLAB codes both in scalar and vector forms on following functions**

A. $\sin^3 x \cos^5 x$

B. $2 + \ln x$

C. $x^4 + 3x - 5$

D. $\dfrac{x^3 - 5}{x^2 - 7x - 7}$

E. $\sqrt{|x^3| + \sec^{-1} x}$

F. $(1 + e^{\sin x})^{x^2 + 3}$

G. $\dfrac{\cosh x + 3}{\sqrt{\dfrac{x+4}{\log_{10}(x^3 - 6)}}}$

H. $\dfrac{1}{(x-3)(x+4)(x-2)}$

I. $\dfrac{1}{1 + \dfrac{1}{1 + \dfrac{1}{x}}}$

J. $\dfrac{a}{x+a} + \dfrac{b}{y+b} + \dfrac{c}{z+c}$

K. $\dfrac{u^2 v^3 w^9}{x^4 y^7 z^6}$

In tabular form, they are coded as follows:

Example	String for scalar computation	String for vector computation
A	sin(x).^3.*cos(x).^5	sin(x)^3*cos(x)^5
B	2+log(x)	2+log(x)
C	x.^4+3*x-5	x^4+3*x-5
D	(x.^3-5)./(x.^2-7*x-7)	(x^3-5)/(x^2-7*x-7)
E	sqrt(abs(x.^3)+asec(x))	sqrt(abs(x^3)+asec(x))
F	(1+exp(sin(x))).^(x.^2+3)	(1+exp(sin(x)))^(x^2+3)
G	(cosh(x)+3)./sqrt((x+4)./log10(x.^3-6))	(cosh(x)+3)/sqrt((x+4)/log10(x^3-6))
H	1./(x-3)./(x+4)./(x-2)	1/(x-3)/(x+4)/(x-2)
I	1./(1+1./(1+1./x))	1/(1+1/(1+1/x))
J	a./(x+a)+b./(y+b)+c./(z+c)	a/(x+a)+b/(y+b)+c/(z+c)
K	u.^2.*v.^3.*w.^9./x.^4./y.^7./z.^6	u^2*v^3*w^9/x^4/y^7/z^6

Finite difference programming in 2D or 3D circumstance dictates the type of code - whether scalar or vector should be employed.

Table A.1   Some mathematical functions and their MATLAB counterparts

Mathematical notation	MATLAB notation	Mathematical notation	MATLAB notation	Mathematical notation	MATLAB notation
$\sin x$	sin(x)	$\sin^{-1} x$	asin(x)	π	pi
$\cos x$	cos(x)	$\cos^{-1} x$	acos(x)	$A+B$	A+B
$\tan x$	tan(x)	$\tan^{-1} x$	atan(x)	$A-B$	A-B
$\cot x$	cot(x)	$\cot^{-1} x$	acot(x)	$A\times B$	A*B
$\cos ecx$	csc(x)	$\sec^{-1} x$	asec(x)	e^x	exp(x)
$\sec x$	sec(x)	$\cos ec^{-1}x$	acsc(x)	A^B	A^B
$\sinh x$	sinh(x)	$\sinh^{-1} x$	asinh(x)	$\ln x$	log(x)
$\cosh x$	cosh(x)	$\cosh^{-1} x$	acosh(x)	$\log_{10} x$	log10(x)
$\sec hx$	sech(x)	$\sec h^{-1}x$	asech(x)	$\log_2 x$	log2(x)
$\cos echx$	csch(x)	$\cos ech^{-1}x$	acsch(x)	Σ	sum
$\tanh x$	tanh(x)	$\tanh^{-1} x$	atanh(x)	Π	prod
$\coth x$	coth(x)	$\coth^{-1} x$	acoth(x)	$\mid x \mid$	abs(x)
10^A	1e A e.g. 1e3	10^{-A}	1e- A e.g. 1e-3	$\sqrt{x}$	sqrt(x)

* In the six trigonometric functions for example sin(x), the x is in radian. If the x is in degree, we use sind(x). The other five functions also have the syntax cosd(x), tand(x), cotd(x), cscd(x), and secd(x) when the x is in degree. The default return from asin(x) is in radian, if you need the return to be in degree, use the command asind(x). Similar degree return is also possible from acosd(x), atand(x), acotd(x), asecd(x), and acscd(x).

Numerical examples to point out the difference between scalar and vector computations are in the following.

We have the matrices $A = \begin{bmatrix} 3 & 5 \\ 7 & 8 \end{bmatrix}$, $B = \begin{bmatrix} 5 & 2 & 1 \\ 0 & 1 & 7 \end{bmatrix}$, and $C = \begin{bmatrix} 3 & 2 & 9 \\ 4 & 0 & 2 \end{bmatrix}$. The scalar computation is not possible between the matrices A and B because of their unequal order, nor is between the matrices A and C for the same reason. On the contrary the scalar multiplication can be conducted between the B and C for having the same order, which is $B.*C = \begin{bmatrix} 15 & 4 & 9 \\ 0 & 0 & 14 \end{bmatrix}$ (element by element multiplication).

Matrix algebra rule says that any matrix A of order $M \times N$ can only be multiplied with another matrix B of order $N \times P$ so that the resulting matrix has the order $M \times P$. For the last paragraph cited A and B, we have $M = 2$, $N = 2$, and $P = 3$ and obtain the vector-multiplied matrix as $A \times B = \begin{bmatrix} 3\times5+5\times0 & 3\times2+5\times1 & 3\times1+5\times7 \\ 7\times5+8\times0 & 7\times2+1\times8 & 7\times1+8\times7 \end{bmatrix} = \begin{bmatrix} 15 & 11 & 38 \\ 35 & 22 & 63 \end{bmatrix}$, which has the MATLAB code A*B not A.*B. Similar interpretation follows for the operators * and /.

Whenever writing the scalar codes A.*B, A./B, and A.^B, we make it certain that both the A and B are identical in matrix size. The 3*A means all elements of matrix A are multiplied by 3 and we do not use 3.*A. Also do we not use A./3 but do A/3. The signs + and - are never preceded by the operator . in scalar codes. The command 4./A means 4 is divided by all elements in A. The A.^4 means power on all elements of A is raised by 4 and so on.

⬥ Scale factors or units

In science and engineering physical quantity measurement always requires the understanding of units. Measured unit of a physical quantity can be far apart from the standard unit very often in the power of 10 that is why scale factors are important. Table A.2 presents the engineering scale factor units and their MATLAB equivalences.

Table A.2 Engineering unit scale factors and their MATLAB counterparts

Scale factor	Symbol	As power of 10	MATLAB code
giga	G	10^9	e9
mega	M	10^6	e6
kilo	K	10^3	e3
milli	m	10^{-3}	e-3
micro	μ	10^{-6}	e-6
nano	n	10^{-9}	e-9
pico	p	10^{-12}	e-12

For example the time $10.7\,m\sec$ is coded as 10.7e-3. Again a distance of $4.7\,km$ is entered by writing 4.7e3 in standard unit.

Appendix B

MATLAB functions/statements for 3D FD study

While working on three dimensional finite difference in MATLAB, we come across many embedded MATLAB functions or programming statements. In order to employ these elements for solving electromagnetic problems, we need to understand their input and output argument types and purpose of the elements. Functions or program elements exercised in the text with brief descriptions are in the following.

B.1 Comparative and logical operators

Comparative operators are used for comparison on two scalar elements, one scalar and one matrix elements, or two identical size matrix elements. There are six comparative operators as presented in table B.1.

Table B.1   Equivalence of comparative operators

Comparative operation	Mathematical notation	MATLAB notation
equal to	$=$	==
not equal to	$\neq$	~=
greater than	$>$	>
greater than or equal to	$\geq$	>=
less than	$<$	<
less than or equal to	$\leq$	<=

The output of expression pertaining to the comparative operators is logical - either true (indicated by 1) or false (indicated by 0). For example when A=3 and B=4, the comparisons A=B, $A \neq B$, A>B, $A \geq B$, A<B, and $A \leq B$ should be false(0), true(1), false(0), false(0), true(1), and true(1)

Table B.2   Scalar comparative operation

>>A=3; B=4; ↵ >>A==B ↵	>>A>B ↵	>>A<B ↵
ans = 0	ans = 0	ans = 1
>>A~=B ↵	>>A>=B ↵	>>A<=B ↵
ans = 1	ans = 0	ans = 1

respectively. We implement these comparative operations as presented in table B.2.

There are two operands A and B in table B.2, each of which is a single scalar. Each of the operands can be a matrix in general. In that case the logical decision takes place element by element on all elements in the matrix. For instance if $A=\begin{bmatrix} 5 & 8 \\ 5 & 7 \end{bmatrix}$ and $B=\begin{bmatrix} 2 & 1 \\ -2 & 9 \end{bmatrix}$, A>B should be $\begin{bmatrix} 5>2 & 8>1 \\ 5>-2 & 7>9 \end{bmatrix} = \begin{bmatrix} 1 & 1 \\ 1 & 0 \end{bmatrix}$. Again if the A happens to be a scalar (say A=4), the single scalar is compared to all elements in the B therefore A≤B should be $\begin{bmatrix} 4 \leq 2 & 4 \leq 1 \\ 4 \leq -2 & 4 \leq 9 \end{bmatrix} =$

$\begin{bmatrix} 0 & 0 \\ 0 & 1 \end{bmatrix}$. In a similar fashion the B also operates on A however the scalar and matrix related comparative implementation is presented in the table B.3.

Some basic logical operations are NOT, OR, and AND. The characters ~, |, and & of the keyboard are adopted for the logical NOT, OR, and AND respectively. In all logical outputs the 1 and 0 stand for true and false respectively. All logical operators apply to the matrices in general. For the matrix $A=\begin{bmatrix} 0 & 0 \\ 0 & 1 \end{bmatrix}$, NOT(A) operation

Table B.3   Scalar and matrix comparative operations

when A and B are matrices,	when A is scalar and B is matrix,
>>A=[5 8;5 7]; ↵	>>A=4; ↵
>>B=[2 1;-2 9]; ↵	>>B=[2 1;-2 9]; ↵
>>A>B ↵	>>A<=B ↵
ans =	ans =
1 1	0 0
1 0	0 1

should provide $\begin{bmatrix} 1 & 1 \\ 1 & 0 \end{bmatrix}$ (see table B.4). The logical OR and AND operations on the like positional elements of the two matrices $A=\begin{bmatrix} 1 & 1 \\ 0 & 1 \end{bmatrix}$ and $B=\begin{bmatrix} 0 & 1 \\ 1 & 1 \end{bmatrix}$ must return $\begin{bmatrix} 1 & 1 \\ 1 & 1 \end{bmatrix}$ and $\begin{bmatrix} 0 & 1 \\ 0 & 1 \end{bmatrix}$ respectively. Table B.4 shows both implementations.

Table B.4   Basic logical operations on matrix elements

for NOT(A) operation,	for A OR B,	for A AND B,	for A XOR B,
>>A=[0 0;0 1]; ↵	>>A=[1 1;0 1]; ↵	>>A&B ↵	>>xor(A,B) ↵
>>~A ↵	>>B=[0 1;1 1]; ↵		
	>>A|B ↵		ans =
		ans =	1 0
ans =	ans =	0 1	1 0
1 1	1 1	0 1	
1 0	1 1		

If the A or the B is a single 1 or 0, it operates on all elements of the other.

Sometimes we need to check the interval of the independent variable of mathematical functions for instance $-6 \le x \le 8$. The interval is split in two parts $-6 \le x$ and $x \le 8$. In terms of the logical statement one expresses the $-6 \le x \le 8$ as (-6<=x)&(x<=8).

There is no operator for the XOR logical operation instead the MATLAB function xor syntaxed by xor(A,B) implements the operation as presented in the table B.4.

B.2 Simple if/if-else/nested if syntax

Conditional commands are exercised by the if-else statements (reserve words). Also comparisons and checkings may need if-else statements. We can have different if-else structures namely simple-if, if-else, or nested-if depending on programming circumstances, some of which we discuss in the following.

⊟ Simple if

The program syntax of simple-if is as follows:

> if *logical expression*
> > *Executable MATLAB command(s)*
>
> end

Logical expression usually requires the use of comparative operators which are explained in appendix B.1. If the logical expression beside the if is true, the command between the if and end is executed otherwise not. In tabular form a simple-if implementation is as follows:

Example: If $x \geq 1$, we compute $y = \sin x$. When $x = 2$, we should see $y = \sin 2 = 0.9093$.	Executable M-file: x=2; if x>=1 y=sin(x); end	Steps: Save the statements in a new M-file (section 1.1) by the name **test** and execute the following: >>test ↵	Check from the command window after running the M-file: >>y ↵ y = 0.9093

⊟ If-else

General program syntax for the if-else structure is as follows:

> if *logical expression*
> > *Executable MATLAB command(s)*
>
> else
> > *Executable MATLAB command(s)*
>
> end

If the logical expression beside the if is true, the command between if and else is executed else the command between else and end is executed. In tabular form, an if-else-end implementation is the following:

Example: When $x = 1$, we compute $y = \sin\dfrac{x\pi}{2} = 1$ otherwise $y = \cos\dfrac{x\pi}{2} = 0$.	Executable M-file: x=1; if x==1 y=sin(x*pi/2); else y=cos(x*pi/2); end	Steps: Save the statements in a new M-file by the name **test** and execute the following: >>test ↵	Check from the command window after running the M-file: >>y ↵ y = 1

If we had x=2; in the first line of M-file in last exercise, we would see y= $\cos\pi$ =−1.

⊟ **Nested-if**

The third type of if structure is the nested-if whose general program syntax is attached in the right side text box. Clearly the syntax takes care of multiple logical expressions which we demonstrate by one example as shown in the following table.

```
if logical expression
       Executable MATLAB command(s)
elseif logical expression
       Executable MATLAB command(s)
       ⋮
elseif logical expression
       Executable MATLAB command(s)
else
       Executable MATLAB command(s)
end
```

| Example: The best example can be taking the decision of grades out of 100 based on the achieved number of a student. The grading policy is stated as if the achieved number of a student is greater than or equal 90, greater than or equal to 80 but less than 90, greater than or equal to 70 but less than 80, greater than or equal to 60 but less than 70, greater than or equal to 50 but less than 60, and less than 50, then the grade is decided as A, B, C, D, E, and F respectively. | Executable M-file:

N=77;
if N>=90
  g='A';
elseif (N<90)&(N>=80)
  g='B';
elseif (N<80)&(N>=70)
  g='C';
elseif (N<70)&(N>=60)
  g='D';
elseif (N<60)&(N>=50)
  g='E';
else
  g='F';
end | In the executable M-file, the N and g refer to the number achieved and the grade respectively. If the number N is 77, the grade g should be C. Any character is argumented under the single inverted comma.

Steps: Save the left statements in a new M-file by the name test and execute the following:
>>test ↵ | Check from the command window after running the M-file:
>>g ↵

g =

C |

B.3 Data accumulation

Sometimes it is necessary that we perform appending operation on an existing matrix at MATLAB workspace.

⬥ **Appending rows**

Assume that the $A = \begin{bmatrix} 1 & 3 & 5 \\ 2 & 6 & 8 \\ 9 & 5 & 0 \\ 4 & 7 & 8 \end{bmatrix}$ is formed by appending two

row matrices [9 5 0] and [4 7 8] with the matrix $B = \begin{bmatrix} 1 & 3 & 5 \\ 2 & 6 & 8 \end{bmatrix}$.

We first enter the matrix B (section 1.1) into MATLAB and append one row after another by using the command as presented below:

for entering B, >>B=[1 3 5;2 6 8] ↵	for appending the first row, >>B=[B;[9 5 0]] ↵	for appending the second row, >>A=[B;[4 7 8]] ↵

B =
 	
    1   3   5
    2   6   8
	B =
	
	    1 3 5
	    2 6 8
	    9 5 0
		A =
		
		    1 3 5
		    2 6 8
		    9 5 0
		    4 7 8

The command B=[B;[9 5 0]] in above execution says that the row [9 5 0] is to be appended with the existing B (inside the third bracket) and that the result is again assigned to B. You can append as many rows as you want. The important point is the number of elements in each row that is to be appended must be equal to the number of columns in the matrix B.

✦ Appending columns

Suppose $C = \begin{bmatrix} 1 & 3 & 5 & 9 & 3 \\ 2 & 6 & 8 & 0 & 1 \\ 9 & 5 & 0 & 1 & 9 \end{bmatrix}$ is formed by appending two

column matrices $\begin{bmatrix} 9 \\ 0 \\ 1 \end{bmatrix}$ and $\begin{bmatrix} 3 \\ 1 \\ 9 \end{bmatrix}$ with matrix $D = \begin{bmatrix} 1 & 3 & 5 \\ 2 & 6 & 8 \\ 9 & 5 & 0 \end{bmatrix}$. We get the

matrix D into MATLAB and append one column after another as follows:

for entering D, >>D=[1 3 5;2 6 8;9 5 0] ↵	for appending the first column, >>D=[D [9 0 1]'] ↵	for appending the second column, >>C=[D [3 1 9]'] ↵

D =
 	
    1   3   5
    2   6   8
    9   5   0
	D =
	
	    1 3 5 9
	    2 6 8 0
	    9 5 0 1
		C =
		
		    1 3 5 9 3
		    2 6 8 0 1
		    9 5 0 1 9

The column matrix [9 0 1]' and D in above execution have one space gap within the third bracket. In the second of above implementation, the resultant matrix is again assigned to D. Append as many columns as you want just remember that the number of elements in each column that is to be appended must be equal to the number of rows in the matrix D.

✦ Data accumulation by using the two appending techniques

Suppose initially there is nothing in the f matrix, which in MATLAB we write by the statement f=[ ]; (an empty matrix is assigned to f). An empty matrix does not have any size and completely empty, it follows the null symbol $\varnothing$ of matrix algebra. Let us say k=2 and perform the assignment as follows:

```
>>f=[ ]; k=2; ↵
```

Now if we execute f=[f k] time and again first f=[f k] returns 2, second f=[f k] returns [2 2], third f=[f k] returns [2 2 2], and so on. This is called row directed data accumulation. Column directed data accumulation occurs by executing f=[f;k] each time.

The demonstrated k is just a scalar but it can be a return from some function, scalar, row matrix, column matrix, or rectangular matrix.

B.4 For-loop syntax

A for-loop performs similar operations for a specific number of times and must be started with the **for** and terminated by an **end** statements. Following the **for** there must be a counter. The counter of the for-loop can be any variable that counts integer or fractional values depending on the increment or decrement. If the MATLAB command statements between the **for** and **end** of a for-loop are few words lengthy, one can even write the whole for-loop in one line. The programming syntax and some examples on the for-loop are as follows:

✦ Program syntax

> for *counter*=starting value:increment or decrement of the
> counter value:final value
> *Executable MATLAB command(s)*
>
> end

✦ Example 1

Our problem statement is to compute $y = \cos x$ for $x = 10^0$ to 70^0 with the increment 10^0. Let us assign the computed values to some variable y where y should be [$\cos 10^0$ $\cos 20^0$ $\cos 30^0$ $\cos 40^0$ $\cos 50^0$ $\cos 60^0$ $\cos 70^0$]=[0.9848   0.9397   0.866   0.766   0.6428   0.5   0.342].

In the programming context, y(1) means the first element in the row matrix y, y(2) means the second element in the row matrix y, and so on. MATLAB code for the $\cos x$ is cosd(x) where x is in degree. The for-loop counter expression should be k=1:1:7 or k=1:7 to have the control on the position index in the row matrix y (because there are 7 elements or indexes in y). Since the computation needs 10 to 70, one generates that by writing k*10. Following is the implementation:

Executable M-file:	*Or, as a one line:*
for k=1:1:7 y(k)=cosd(k*10); end	for k=1:1:7 y(k)=cosd(k*10); end

Steps we need:

Open a new M-file (section 1.1), type the executable M-file statements in the M-file editor, save the editor contents by the name **test** in your working path, and call the **test** as shown below.

```
>>test ↵
>>y ↵
```

y =

        0.9848 0.9397 0.8660 0.7660   0.6428 0.5000   0.3420

✦ Example 2

A for-loop helps us accumulate data (appendix B.3) controlled by the consecutive loop index. In this example we accumulate some data row directionally according to the for-loop counter index.

For k =1, 2, and 3, we intend to accumulate the k^2 side by side. At the end we should be having [1   4   9] assigned to some variable f – this is our problem statement.

for the right shifting,	for the left shifting,
>>f=[ ]; for k=1:3 f=[f k^2]; end ↵	>>f=[ ]; for k=1:3 f=[k^2 f]; end ↵
>>f ↵	>>f ↵
f =	f =
1 4 9	9 4 1

The for-loop for the accumulation is presented above. The accumulation may occur as right or left shifting. Corresponding to the right shifting, the vector code (appendix A) for k^2 is k^2. The statement f=[ ]; means that an empty matrix is assigned to f outside the for-loop but at the beginning. The k variation in our problem is put as the for-loop counter. How the for-loop accumulates is shown below:

When k=1, f=[f k^2]; returns f=[[ ] 1^2]; ⇒ f=1;
When k=2, f=[f k^2]; returns f=[1 2^2]; ⇒ f=[1 4];
When k=3, f=[f k^2]; returns f=[1 4 3^2]; ⇒ f=[1 4 9];

The accumulation is happening from the left to the right. A single change provides the shifting from the right to the left which is f=[k^2 f];. The complete code and its execution result are also shown above by the heading 'for the left shifting'.

✦ Example 3

Another accumulation can be column directed that is we wish to see the output like $\begin{bmatrix} 1 \\ 4 \\ 9 \end{bmatrix}$ in example 2.

We just insert the row separator of a rectangular matrix (done by the operator ;) in the command f=[f k^2];. Again the shifting can happen either from the up to down or from the down to up. Both implementations are shown below:

for the down shifting,
```
>>f=[ ]; for k=1:3 f=[f;k^2]; end ↵
>>f ↵
```

f =

    1
    4
    9

for the up shifting,
```
>>f=[ ]; for k=1:3 f=[k^2;f]; end ↵
>>f ↵
```

f =

    9
    4
    1

✦ Example 4

Many finite difference problems need writing multiple for-loops. Usually one loop is for one dimensional function, two loops are for two dimensional function, and so on. One dimensional function data takes the form of a row or column matrix.

Suppose we have the one dimensional data as $y = [9\ 6\ 7\ 4\ 6]$. We wish to access to every data in y. A single for-loop helps us conduct that as shown below:

```
>>y=[9 6 7 4 6]; for k=1:length(y) v=y(k); end ↵
```

First we assign the data to workspace y as a row matrix. The command **length** finds the number of elements in the row matrix y. The y(k) means the k-th element in the y which we assign to workspace v (any user-chosen variable). Every single data of the y is available sequentially in the v. The contents of y can be a column matrix too.

B.5 Finding the maximum/minimum numerically

Given a matrix, one finds the maximum element from the matrix by using the command **max** (min for the minimum). Let us say we have three matrices $R = [1\ -2\ 3\ 9]$, $C = \begin{bmatrix} 23 \\ -20 \\ 30 \\ 8 \end{bmatrix}$, and $A = \begin{bmatrix} 2 & 4 & 7 \\ -2 & 7 & 9 \\ 3 & 8 & -8 \end{bmatrix}$ whose maxima are 9, 30, and 9 (from all elements in the matrix) and minima are -2, -20, and -8 respectively. We find the maxima first entering (section 1.1) the respective matrices as follows:

for the row matrix,
```
>>R=[1 -2 3 9]; ↵
>>max(R) ↵
```

ans =

    9
```
>>min(R) ↵
```

ans =

    -2

for the column matrix,
```
>>C=[23;-20;30;8]; ↵
>>max(C) ↵
```

ans =

    30
```
>>min(C) ↵
```

ans =

    -20

for the rectangular matrix,
```
>>A=[2 4 7;-2 7 9;3 8 -8]; ↵
>>max(max(A)) ↵
```

ans =

    9
```
>>min(min(A)) ↵
```

ans =

    -8

Font equivalence is maintained by using the same letter for example A$\Leftrightarrow$ *A* in last implementation. If the matrix is a row or column one, we apply one max or min. For a rectangular matrix, the max or min separately operates on each column that is why two max or min functions are required. The functions are equally applicable on decimal number elements.

In the row matrix *R*, the maximum 9 is occurring as the fourth element in the matrix. Suppose we intend to find the position index (that is 4) of the maximum element in the *R*. Now we need two output arguments – one for the maximum and the other for its index. Its implementation is shown in the right side attached text box of this paragraph in which the two output arguments M and I correspond to the maximum and its integer index respectively.

The function min keeps this type of integer index returning option in a similar fashion.

```
for index finding in R,
>>[M,I]=max(R) ↵

M =
        9
I =
        4
```

B.6 Matrix data flipping

We flip any matrix data by embedded function fliplr or flipud depending on the direction about which it is to be flipped.

Flipping from left to right:

Flipping from left to right of a row or rectangular matrix is performed by function fliplr (abbreviation for flipping from left to right). Suppose we have row matrix $R=[2 \quad 4 \quad 3 \quad -4 \quad 6 \quad 9 \quad 3 \quad 7 \quad 10]$. If you flip the elements of *R* from left to right, in the resulting matrix OR, 10 (last element of *R*) comes first, 7 comes second (second element of *R* from last), 3 comes third (third element of *R* from last), and so do the others hence OR= [10 7 3 9 6 –4 3 4 2].

For a rectangular matrix, flipping operation from left to right will be over each column. That means the first column will be the last column and the second column will be the second from last, and so will be the other columns. Assume that we have rectangular matrix $A = \begin{bmatrix} 4 & 23 & 85 & 34 \\ 5 & 43 & 41 & 87 \\ 8 & 65 & 76 & 71 \end{bmatrix}$.

Flipping from left to right of *A* gives us OA where OA= $\begin{bmatrix} 34 & 85 & 23 & 4 \\ 87 & 41 & 43 & 5 \\ 71 & 76 & 65 & 8 \end{bmatrix}$. Since column matrices have only one column, you will not see any change to a column matrix brought about by fliplr. Have the implementation as shown below:

for the row matrix,
```
>>R=[2 4 3 -4 6 9 3 7 10]; ↵
>>OR=fliplr(R) ↵

OR =
```

$$10 \quad 7 \quad 3 \quad 9 \quad 6 \quad -4 \quad 3 \quad 4 \quad 2$$

for rectangular matrix,
```
>>A=[4 23 85 34;5 43 41 87;8 65 76 71]; ↵
>>OA=fliplr(A) ↵
```

OA =

$$\begin{array}{cccc} 34 & 85 & 23 & 4 \\ 87 & 41 & 43 & 5 \\ 71 & 76 & 65 & 8 \end{array}$$

Flipping from up to down:

The flipud (abbreviation of <u>fl</u>ipping from <u>u</u>p to <u>d</u>own) flips elements of a column or rectangular matrix from up to down. Flip the column matrix $C=\begin{bmatrix}4\\7\\8\\3\\1\end{bmatrix}$ from up to down to get the resulting matrix $OC=\begin{bmatrix}1\\3\\8\\7\\4\end{bmatrix}$. Flipping operation from up to down of a rectangular matrix will happen over each row. Assume that we have rectangular matrix $D=\begin{bmatrix}4 & 23 & 85\\5 & 43 & 41\\8 & 65 & 9\\3 & 12 & 13\end{bmatrix}$. Flip D from up to down to have $OD=\begin{bmatrix}3 & 12 & 13\\8 & 65 & 9\\5 & 43 & 41\\4 & 23 & 85\end{bmatrix}$. No change occurs to a row matrix due to the use of flipud for the reason that row matrices have only one row. Both examples are implemented as follows:

for column matrix,
```
>>C=[4 7 8 3 1]'; ↵
>>OC=flipud(C) ↵
```

OC =

$$\begin{array}{c} 1 \\ 3 \\ 8 \\ 7 \\ 4 \end{array}$$

for rectangular matrix,
```
>>D=[4 23 85;5 43 41;8 65 9;3 12 13]; ↵
>>OD=flipud(D) ↵
```

OD =

$$\begin{array}{ccc} 3 & 12 & 13 \\ 8 & 65 & 9 \\ 5 & 43 & 41 \\ 4 & 23 & 85 \end{array}$$

B.7 Three dimensional array

Matrix-oriented arrangement of data is not convenient for multidimensional and group-related problems. There are many data types exercised in MATLAB, one of which is the three dimensional array. A rectangular matrix has two dimensions. The reason we say two dimensions is any element position of the rectangular matrix needs two indexes to describe it - row and column. For example a 4×4 rectangular matrix has the following position indexes:

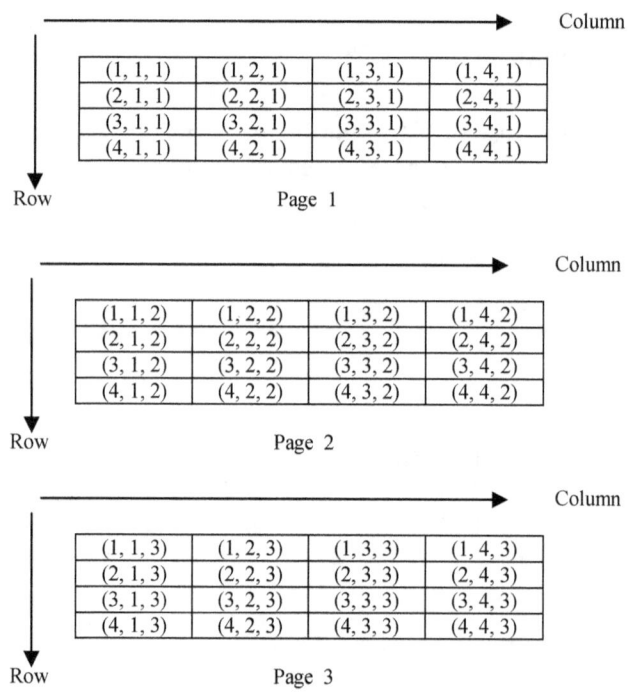

Suppose the above 4×4 rectangular matrix fits in one page of a book. If we have two more pages each containing a 4×4 rectangular matrix, how can one accommodate the three pages in one variable? This necessitates the use of a three dimensional array. The position indexes of the rectangular matrices contained in the three pages can be labeled as follows:

If one gathers the three pages one after another, the three dimensional block of figure 1.B is formed that is how a three dimensional array is created. One can assign any integer or floating-point values to these position indexes.

There are three position indexes of an element in the three dimensional array - dimension 1 (row), dimension 2 (column), and dimension 3 (page).

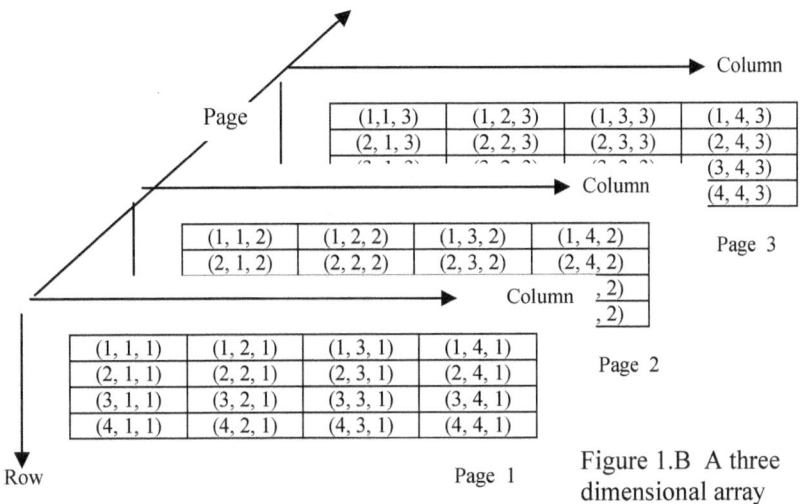

Figure 1.B A three dimensional array

Now we enter the data of a three dimensional array. Assume that the three page data is given as follows; page 1: $\begin{bmatrix} 8 & 3 & 6 \\ 2 & 2 & 1 \end{bmatrix}$, page 2: $\begin{bmatrix} 0 & 4 & 4 \\ 5 & 3 & 8 \end{bmatrix}$, and page 3: $\begin{bmatrix} -1 & 2 & 7 \\ -5 & 5 & 6 \end{bmatrix}$. Name the three dimensional array as A and following is the implementation:

```
>>A(:,:,1)=[8 3 6;2 2 1];      ← Enter the elements of page 1
>>A(:,:,2)=[0 4 4;5 3 8];      ← Enter the elements of page 2
>>A(:,:,3)=[-1 2 7;-5 5 6];    ← Enter the elements of page 3
>>A                            ← To see what in A

A(:,:,1) =                     ← Corresponds to page 1
        8   3   6
        2   2   1
A(:,:,2) =                     ← Corresponds to page 2
        0   4   4
        5   3   8
A(:,:,3) =                     ← Corresponds to page 3
       -1   2   7
       -5   5   6
```

Most manipulations of a rectangular matrix can be extended for any three dimensional array. Some manipulations pertaining to A are presented in the following. The third page element 7 has the index (1,3) that is called by:

```
>>A(1,3,3)
```

ans =
       7

You can change the value say by 10 and have it as follows:

    >>A(1,3,3)=10; ↲
    >>A(:,:,3) ↲   ← To see only the third page

ans =

    -1   2 10
    -5   5   6

Suppose you want to remove the third page from A hence carry out the following:

    >>A(:,:,3)=[ ]; ↲

Now A is containing the first two pages. Squares of all elements of A are $\begin{bmatrix} 64 & 9 & 36 \\ 4 & 4 & 1 \end{bmatrix}$ (page 1) and $\begin{bmatrix} 0 & 16 & 16 \\ 25 & 9 & 64 \end{bmatrix}$ (page 2) respectively. You can have that on execution of A.^2 as follows:

    >>A.^2 ↲

    ans(:,:,1) =
            64 9 36
             4 4   1
    ans(:,:,2) =
             0 16 16
           25   9 64

Again a three dimensional array A can be converted to a column matrix by using the command A(:).

A long row or column matrix can be converted to a three dimensional array by using the command **reshape**. For example choose R= [1 8   61 11   40   68   34   12   45   32   89   43]. There are twelve elements in R. The product of row, column, and page numbers must be 12 and the formation is shown below:

    >>R=[1 8 61 11 40 68 34 12 45 32 89 43]; ↲
    >>reshape(R,2,3,2) ↲

    ans(:,:,1) =   ← Displays the first page
            1 61 40
            8 11 68
    ans(:,:,2) =   ← Displays the second page
          34 45 89
          12 32 43

Note that the conversion is in accordance with column.

One can add one more page by assigning a 2×3 matrix to A(:,:,3).

Adding some scalar say 5 to each element of A is accomplished by A+5.

Multiplication of a three dimensional array is not defined. Pages of the array can be multiplied according to the rules of matrix algebra for example page 2 with page 3 by A(:,:,2)*A(:,:,3). For simplicity we have shown all manipulations taking the integer elements but the elements can be floating-points, characters, or even symbolic variables.

B.8 Array padding by user-supplied elements

Suppose we have the matrix $A = \begin{bmatrix} 11 & 10 & 11 \\ 12 & 10 & -2 \end{bmatrix}$. We wish to pad the

A in different directions by some element 30. Enter the matrix to **A** by:

```
>>A=[11 10 11;12 10 -2]; ↵
```

The embedded function we apply is **padarray** which uses a syntax new array name=**padarray**(original array, times to be repeated the element into, element, reserve word under quote for direction). Before, after, and all around are indicated by the reserve words **pre**, **post**, and **both** respectively.

To get $\begin{bmatrix} 30 & 30 & 30 \\ 11 & 10 & 11 \\ 12 & 10 & -2 \end{bmatrix}$ from A, we pad before one time in the

column direction:

```
>>B=padarray(A,1,30,'pre') ↵


B =
        30   30   30
        11   10   11
        12   10   -2
```

In the same direction two times pre padding we get by:

```
>>B=padarray(A,2,30,'pre') ↵


B =
        30   30   30
        30   30   30
        11   10   11
        12   10   -2
```

In the same direction two times post padding we get by:

```
>>B=padarray(A,2,30,'post') ↵


B =
        11   10   11
        12   10   -2
        30   30   30
        30   30   30
```

The second input argument can be a two element row matrix too (that is dimension of the matrix to be padded) in that case padding takes place as matrices. Call it for 2×2 size all around for ongoing example:

```
>>B=padarray(A,[2 2],30,'both') ↵


B =
        30   30   30   30   30   30   30
        30   30   30   30   30   30   30
        30   30   11   10   11   30   30
        30   30   12   10   -2   30   30
        30   30   30   30   30   30   30
        30   30   30   30   30   30   30
```

As you see in the last return the element 30 is all around the A. What if we need the padding in the row direction? The first element of the second input argument is 0 then. Call the function for left padding in the row direction by:

```
>>B=padarray(A,[0 1],30,'pre') ↵
```

B =

$$\begin{array}{cccc} 30 & 11 & 10 & 11 \\ 30 & 12 & 10 & -2 \end{array}$$

Execute **help padarray** for more options.

B.9 Some operators to handle finite difference

MATLAB keeps provision for embedded operators facilitating FD computing which are applied for summation, differentiation, and integration. The issue of differentiation is rigorously addressed in earlier chapters given the nature of 3D FD. We wish to address summation and integration using FD in this appendix.

✦ Summation operator

MATLAB function **sum** adds all elements in a row, column, or rectangular matrix when the matrix is its input argument. Example matrices are $R=[1 \ -2 \ 3 \ 9]$, $C=\begin{bmatrix} 23 \\ -20 \\ 30 \\ 8 \end{bmatrix}$, and $A=\begin{bmatrix} 2 & 4 & 7 \\ -2 & 7 & 9 \\ 3 & 8 & -8 \end{bmatrix}$ whose all element sums

are 11, 41, and 30 for the R, C, and A respectively. We execute the summations as follows (font equivalence is maintained by using the same letter for example A$\Leftrightarrow A$):

Sum for the row matrix,	Sum for the column matrix,	Sum for the rectangular matrix,
>>R=[1 -2 3 9]; ↵	>>C=[23 -20 30 8]'; ↵	>>A=[2 4 7;-2 7 9;3 8 -8]; ↵
>>sum(R) ↵	>>sum(C) ↵	>>sum(sum(A)) ↵
		ans =
ans =	ans =	30
11	41	

For a rectangular matrix, two functions are required because the inner sum performs addition over each column and the result is a row matrix. The outer sum provides addition over the resulting row matrix. The function is operational on real, complex, and even symbolic variable like x or y. Just two examples are furnished below.

Example 1:

Suppose $\sum_{p=-2}^{2} p^2 = 10$ we intend to calculate.

We generate the p values as a row or column matrix (section 1.3), write scalar code of p^2 for computing (appendix A), and use the command **sum** as follows:

```
>>p=-2:2; ⌋   ← p⇔ p , p is user-chosen, holds values as a row matrix
>>f=p.^2; ⌋   ← f is user-chosen, holds computed values as a row matrix
>>sum(f) ⌋
```

```
ans =
        10
```

Example 2:

Not necessarily the p should be consecutive say $\sum pf[p]=60$

is to be implemented where $p=[1 \quad 4 \quad 9]$ and $f[p]=[-20 \quad 2 \quad 8]$.
Execute the following for the computation:

```
>>p=[1 4 9]; ⌋   ← p⇔ p , p is user-chosen, holds values as a row matrix
>>f=[-20 2 8]; ⌋ ← f⇔ f[p], user-chosen, holds values as a row matrix
>>sum(p.*f) ⌋    ← .* is used for element by element multiplication
```

```
ans =
        60
```

The p can be even fractional or complex in last examples.

✦ Integral operator

There is no hard and fast integral operator for FD because integration is valid for continuous functions. The **sum** implements the $\int .....$ Say we have

to compute $\int_{x_1}^{x_2} f(x)dx$ by using the finite difference technique. Obviously the

x interval is $x_1 \le x \le x_2$. User has to decide resolution Δx or number of samples N over the interval. Once known, the FD computing for the integration is $\Delta x \sum_m f(m\Delta x)$. First generate all x samples as a row or column vector, then use the scalar code (appendix A) to get $f(m\Delta x)$ values, and next apply the command **sum**. Let us see the following examples in this regard.

Example 1:

It is given that $\int_{x=-1}^{x=5}(-4x^2+8x-9)dx=-126$. Implement the

integration by the FD method with 301 samples.

Clearly we have $N=301$, $x_1=-1$, and $x_2=5$ so $\Delta x=\dfrac{x_2-x_1}{N-1}$

and execute the following:

```
>>N=301; ⌋              ← N⇔ N , N is user-chosen
>>dx=(5-(-1))/(N-1); ⌋  ← dx⇔ Δx , dx is user-chosen
>>x=-1:dx:5; ⌋          ← x holds x  samples as a row matrix, x user-chosen
>>f=-4*x.^2+8*x-9; ⌋    ← f holds f(mΔx) or f(x) samples as a row
                          matrix, f is user-chosen
```

>>dx*sum(f) ⏎   ← Exercising $\Delta x \sum_m f(m\Delta x)$

ans =
          -126.9016

Example 2:

Think about $4\int_{x=-1}^{x=5} dx = 24$, how is it different from the example 1? No definitive x related expression is there in the integrand. If we choose 301 samples over $-1 \le x \le 5$, the value of each sample is constant or 4. We first generate (appendix B.10) necessary number of ones and then multiply by 4 to get the integrand samples, following is the complete execution:

>>N=301; ⏎   ← N⇔ N
>>dx=(5-(-1))/(N-1); ⏎   ← dx⇔ Δx
>>x=-1:dx:5; ⏎   ← x holds x samples as a row matrix
>>f=4*ones(1,N); ⏎   ← f holds $f(x)$ samples as a row matrix

>>dx*sum(f) ⏎   ← Exercising $\Delta x \sum_m f(m\Delta x)$

ans =
          24.0800

Discrepancy and exception:

In example 2 the correct result is 24 but we found 24.08. Discrepancy is always there while using FD. How much relative error is associated? The answer is (24.08-24)/24×100=0.33% - extremely low indeed. Reduce Δx by taking more sample numbers certainly the result will improve.

B.10 Matrix of ones, zeroes, and constants

MATLAB built-in commands **ones** and **zeros** implement user-defined matrix of ones and zeroes respectively. Each function conceives two input arguments, the first and second of which are the required numbers of rows and columns respectively. Let us say we intend to form the matrices A

$= \begin{bmatrix} 1 & 1 & 1 \\ 1 & 1 & 1 \\ 1 & 1 & 1 \\ 1 & 1 & 1 \end{bmatrix}$, $B = \begin{bmatrix} 1 & 1 & 1 \\ 1 & 1 & 1 \\ 1 & 1 & 1 \end{bmatrix}$, and $C = \begin{bmatrix} 1 & 1 & 1 & 1 \\ 1 & 1 & 1 & 1 \end{bmatrix}$. Their orders are 4×3, 3×3, and 2×4

respectively and the implementations are as follows:

for A,   for B,   for C,
>>A=ones(4,3) ⏎   >>B=ones(3) ⏎   >>C=ones(2,4) ⏎

A =   B =   C =
    1 1 1   1 1 1   1 1 1 1
    1 1 1   1 1 1   1 1 1 1
    1 1 1   1 1 1
    1 1 1

Either the number of rows or columns will do if the matrix is a square. For the row and column matrices of ones for example of length 6, the commands would be **ones(1,6)** and **ones(6,1)** respectively.

Formation of the matrix of zeroes is quite similar to that of the matrix of ones. Replacing the function **ones** by **zeros** does the formation.

Matrix of zeroes like $A = \begin{bmatrix} 0 & 0 & 0 \\ 0 & 0 & 0 \\ 0 & 0 & 0 \\ 0 & 0 & 0 \end{bmatrix}$, $B = \begin{bmatrix} 0 & 0 & 0 \\ 0 & 0 & 0 \\ 0 & 0 & 0 \end{bmatrix}$, and $C = \begin{bmatrix} 0 & 0 & 0 & 0 \\ 0 & 0 & 0 & 0 \end{bmatrix}$ (whose

orders are 4×3, 3×3, and 2×4) we form by the commands **A=zeros(4,3)**, **B=zeros(3)**, and **C=zeros(2,4)** respectively. A row and a column matrices of 6 zeroes are formed by the commands **zeros(1,6)** and **zeros(6,1)** respectively.

A matrix of constants is obtained by first creating a matrix of ones of the required size and then multiplying by the constant number. For example

the matrix $\begin{bmatrix} 0.2 & 0.2 & 0.2 \\ 0.2 & 0.2 & 0.2 \\ 0.2 & 0.2 & 0.2 \\ 0.2 & 0.2 & 0.2 \end{bmatrix}$ is generated by the command **0.2*ones(4,3)**.

Appendix C

Some graphing functions of MATLAB

One of MATLAB's nicest features is you can have your graphics drawn while programming finite difference related problems. There are so many easy accessible embedded graphics functions that one finds it very interesting when the input-output argumentation style of these functions is understood. Some graphing functions which we applied frequently in previous chapters are addressed for syntax details in the sequel.

◆ y versus x data

The command plot graphs y versus x data. Let us say we have the attached (on the right side in this paragraph) tabular data. We intend to graph these data as y versus x graph.

Tabular data of y versus x type:							Command to graph the y vs x data:
x	-6	-4	0	4	5	7	>>x=[-6 -4 0 4 5 7]; ⏎
y	9	3	-3	-5	2	0	>>y=[9 3 -3 -5 2 0]; ⏎ >>plot(x,y) ⏎

Commands to graph the data are also presented beside the tabular data on the right side in the last paragraph. We first assign the x and y data to workspace x and y (some user-chosen variables) respectively and then call the command plot to see the figure C.1(a). The plot has two input arguments, the first and second of which are the x and y data both as a row or column matrix of identical size respectively.

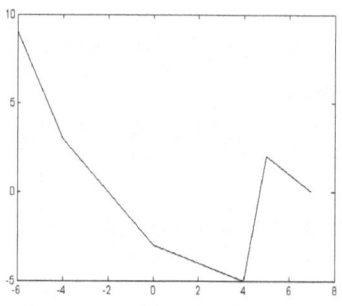

Figure C.1(a) y vs x plot of the tabular data

In order to graph a mathematical expression by using the plot, one first needs to calculate the functional values by using the scalar code (appendix A) and then applies the command. During the calculation, computing step selection is mandatory which is completely user-defined.

For instance we wish to graph the function $f(x) = x^2 - x + 2$ over $-2 \leq x \leq 3$.

Let us choose some x step 0.1. The x vector as a row matrix is generated by x=-2:0.1:3; (section 1.1). At every element in x vector, the functional value is computed and assigned to workspace f by f=x.^2-x+2;. The f is any user-chosen variable. Now we call the grapher as plot(x,f) to see the trajectory (not shown for space reason).

The command **plot** just draws the graph, no graphical features such as x axis label or title are added to the graph. It is the user who is supposed to add these graphical features.

◆ Multiple y data versus common x data

The **plot** keeps many options, one of which is just discussed. We graph several y data versus common x data with the help of **plot** but with different number of input arguments. Let us choose the right side attached table for graphing.

Tabular data for multiple y versus common x :						
x	-6	-4	0	4	5	7
y_1	9	3	-3	-5	2	0
y_2	0	-2	1	0	5	7.7
y_3	-1	2	8	1	0	-3

We intend to plot the y_1, y_2, and y_3 on common x data. To do so,

>>x=[-6 -4 0 4 5 7]; ↵   ← Assigning the x data as a row matrix to x
>>y1=[9 3 -3 -5 2 0]; ↵   ← Assigning the y_1 data as a row matrix to y1
>>y2=[0 -2 1 0 5 7.7]; ↵   ← Assigning the y_2 data as a row matrix to y2
>>y3=[-1 2 8 1 0 -3]; ↵   ← Assigning the y_3 data as a row matrix to y3
>>plot(x,y1,x,y2,x,y3) ↵   ← Applying the command plot

The **plot** now has six input arguments - two for each graph, the first and second of which are the common x and y data to be plotted respectively. If there were four y data, the command would be plot(x,y1,x,y2,x,y3,x,y4). Once the data is plotted for several y, identifying the y traces is obvious which is carried out by the command **legend**. The command legend('y1', 'y2','y3') puts identifying

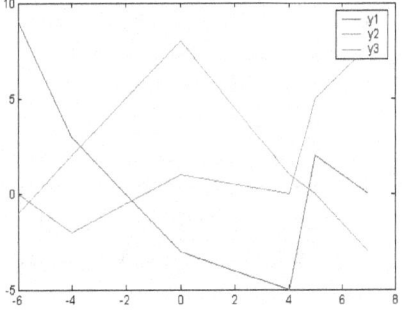

Figure C.1(b) Multiple y vs x for the tabular data

marks/colors among various graphs. The input argument of the **legend** is any user-given word but under quote and separated by a comma. The number of y traces must be equal to the number of input arguments of **legend**. We gave the names y1, y2, and y3 for the three y traces respectively. In doing so, we end up with the figure C.1(b). You can even move the legend on the plot area by using mouse.

You see all graphics throughout the text as black and white because we did not include color graphics in the text (for expense reason). But MATLAB displays figures in color plots, which you can easily identify.

Another situation can be that we have several functions and intend to plot those on common x variation. For instance we wish to graph $y_1 = x^3 - x^2 + 4$ and $y_2 = x^2 - 7x - 5$ over the common $-1 \le x \le 3$.

Under these circumstances, the step selection of x data is compulsory. Without calculating the functional values of given y curves, we can not graph the functions for which we exercise the scalar code. Let us choose the x step as 0.1. We first generate the common x vector as a row matrix by writing x=-1:0.1:3; and then calculate the y_1 and y_2 (y1$\Leftrightarrow y_1$ and y2$\Leftrightarrow y_2$) data by writing y1=x.^3-x.^2+4; y2=x.^2-7*x-5; and eventually the graph appears by executing plot(x,y1,x,y2), graph is not shown for space reason. Thus you can graph three or more functions.

✦ Functions of the form $y = f(x)$

If any function is of the form $y = f(x)$ and the $f(x)$ versus x is to be graphed, the built-in ezplot is the best option which uses a syntax ezplot(functional vector code under quote according to appendix A, interval bounds as a two element row matrix) where the first and second elements in the row matrix are beginning and ending bounds of the interval respectively. The ezplot graphs $y = f(x)$ in the default interval $-2\pi \le x \le 2\pi$ when no interval description is argumented.

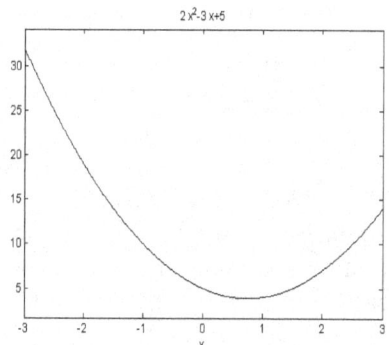

Figure C.1(c) Plot of $y = 2x^2 - 3x + 5$ versus x over $-3 \le x \le 3$

We intend to graph the function $y = 2x^2 - 3x + 5$ over the interval $-3 \le x \le 3$. We first give $2x^2 - 3x + 5$ MATLAB vector code and then assign that to y as follows:
```
>>y='2*x^2-3*x+5'; ↵
```
In above implementation the y is any user-chosen variable. The interval $-3 \le x \le 3$ is entered by [-3 3]. To obtain the plot of y in the given interval, we execute the following at the command prompt:
```
>>ezplot(y,[-3,3]) ↵
```
Above command results the figure C.1(c).

✦ Multiple graphs in the same window

The function subplot splits a figure window in subwindows based on the user definition. It accepts three positive integer numbers as the input arguments, the first and second of which indicate the number of subwindows in the horizontal and the number of subwindows in the vertical directions respectively. For example 22 means two subwindows horizontally and two subwindows vertically, 32 means three subwindows horizontally and two subwindows vertically, ... and so on. The third integer in the input argument numbered consecutively offers control on the subwindows so generated. If the first two digits are 32, there should be 6 subwindows

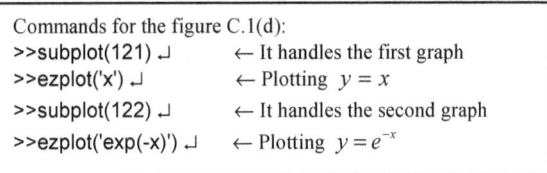

Commands for the figure C.1(d):
>>subplot(121) ↵   ← It handles the first graph
>>ezplot('x') ↵   ← Plotting $y = x$
>>subplot(122) ↵   ← It handles the second graph
>>ezplot('exp(-x)') ↵   ← Plotting $y = e^{-x}$

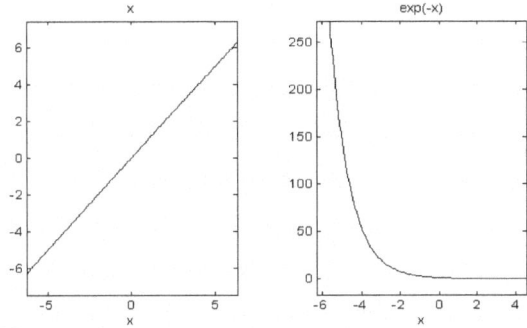

Figure C.1(d) Plots of $y = x$ and $y = e^{-x}$ side by side in the same window

and they are numbered and controlled by using 1 through 6. When you plot some graph in a subwindow, as if you are handling an independent figure window.

We wish to graph $y = x$ and $y = e^{-x}$ side by side as two different plots by using earlier mentioned **ezplot** but in the same window. If we imagine the subfigures as matrix elements, we have a figure matrix of size 1×2 (one row and two columns). That is why the first two integers of the input argument of **subplot** should be 12. Attached commands in the upper right text box of last paragraph show the figure C.1(d). The third integers 1 and 2 in the **subplot** give the control on the first and second subfigures respectively.

As another example we wish to plot $y = x$ and $y = e^{-x}$ in the upper row and only $y = (1 - e^{-x})$ in the lower row subfigures in the same window

Commands for the figure E.1(e):
>>subplot(221) ↵   ← Subfigure selection for $y = x$
>>ezplot('x') ↵   ← Plotting $y - x$
>>subplot(222) ↵   ← Subfigure selection for $y = e^{-x}$
>>ezplot('exp(-x)') ↵   ← Plotting $y = e^{-x}$
>>subplot(212) ↵   ← Subfigure selection for $y = (1 - e^{-x})$
>>ezplot('1-exp(-x)') ↵   ← Plotting $y = (1 - e^{-x})$

whose implementation needs above attached text box commands and whose final output is the figure C.1(e). We are supposed to have four figures when the integer input argument of **subplot** is 22 (two for rows and two for columns). The arguments 221, 222, 223, and 224 provide handle on the four figures consecutively. The figures could have been plotted on 223 and 224 are absent so we ignore them. The argument 21 creates two subfigures (two rows and one column) handled by 211 and 212, but 211 is absent so we ignore that too.

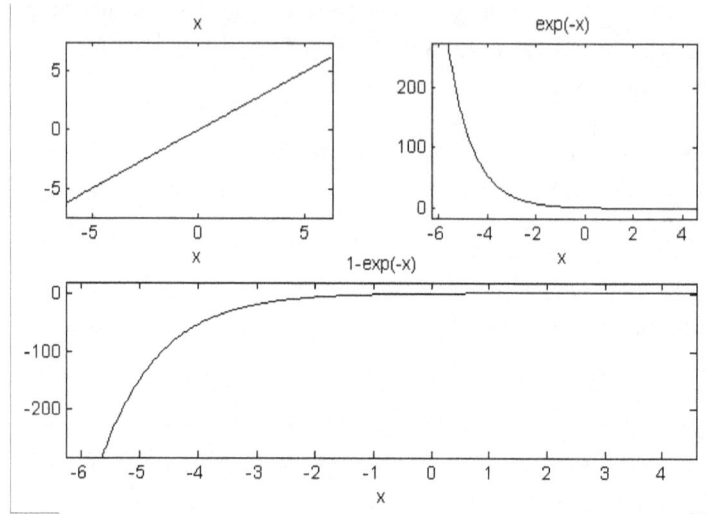

Figure C.1(e) Plots of $y = x$ and $y = e^{-x}$ in the upper row and $y = (1 - e^{-x})$ in the lower row in the same window

Let us see the input arguments of **subplot** for different subfigures (each third brace set [ ] is one subfigure in the following tabular representation) as follows:

Subfigures needed	First two input integers of subplot	Third input integer of subplot	Commands we need
[ ] [ ]	22	[1] [2]	subplot(221) subplot(222)
[ ] [ ]		[3] [4]	subplot(223) subplot(224)
[ ] [ ]	22 for upper two (lower two remain empty)	[1] [2]	subplot(221) subplot(222)
[ ]	21 for the lower one (upper one remains empty)	[2]	subplot(212)
[ ]	21 for the upper one (lower one remains empty)	[1]	subplot(211) subplot(223)
[ ] [ ]	22 for the lower two (upper two remain empty)	[3] [4]	subplot(224)
[ ] ⎡ ⎤	22 for the left two (right two remain empty)	[1] ⎡ ⎤	subplot(221) subplot(223)
[ ] ⎣ ⎦	12 for the right one (left one remains empty)	[3] ⎣ 2 ⎦	subplot(122)
⎡ ⎤ [ ]	22 for right two (left two remain empty)	⎡ ⎤ [2]	subplot(222) subplot(224)
⎣ ⎦ [ ]	12 for the left one (right one remains empty)	⎣ 1 ⎦ [4]	subplot(121)

♦ Symbol on a drawn graph

Let us say $r = e^{-2\theta}\sin 2\theta$ is to be plotted over $0 \leq \theta \leq \pi$. Earlier quoted **ezplot** graphs the function by **ezplot('exp(-2*t)* sin(2*t)',[0 pi]) - t** is used for θ, figure C.1(f) shows it. We would like to drop the text **The plot of** $e^{-2\theta}\sin 2\theta$ on the graph. The command **gtext** gives the provision for dropping a mouse-driven text (written under quote) on a drawn graphics. Execute the following at the command prompt:

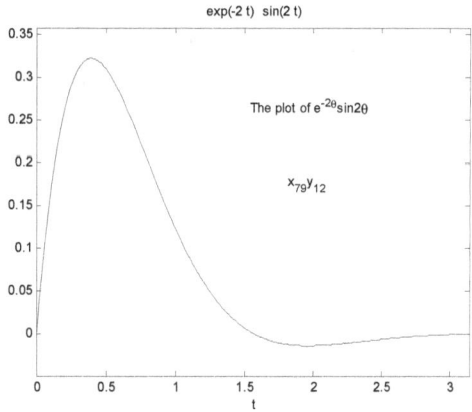

Figure C.1(f) Plot of $r = e^{-2\theta}\sin 2\theta$

```
>>gtext('The plot of e^-^2^\thetasin2\theta') ↵
```

Table C.1   MATLAB codes for various symbols (not in alphabetical order)

MATLAB code	Symbol	MATLAB code	Symbol	MATLAB code	Symbol	MATLAB code	Symbol
\omega	ω	\gamma	γ	\mu	μ	\Xi	Ξ
\Omega	Ω	\Gamma	Γ	\nu	ν	\xi	ξ
\phi	ϕ	\delta	δ	\surd	$\sqrt{\ }$	\oplus	$\oplus$
\Phi	Φ	\Delta	Δ	\in	$\in$	\alpha	α
\zeta	ζ	\epsilon	ε	\chi	χ	\sim	$\sim$
\pi	π	\eta	η	\leq	$\leq$	\iota	ι
\Pi	Π	\Psi	Ψ	\geq	$\geq$	\infty	∞
\beta	β	\psi	ψ	\pm	$\pm$	\exists	$\exists$
\theta	θ	\kappa	κ	\int	$\int$	\cap	$\cap$
\Theta	Θ	\Sigma	Σ	\copyright	$\copyright$	\subset	$\subset$
\lambda	λ	\sigma	σ	\nabla	∇	\ni	$\ni$
\Lambda	Λ	\neq	$\neq$	\upsilon	υ	\oslash	$\oslash$
\partial	∂	\rho	ρ	\tau	τ	\otimes	$\otimes$

Then go to the figure window and you find the mouse pointer activated and a crosshair is appearing. Choose any convenient position in the graph and click the left button of mouse to see the inside text as in figure C.1(f). The symbol θ is written by the command \theta in graphics. Any superscript is placed by the command ^, as explanation we can say $e^{-} \Leftrightarrow$ e^-, $e^{-2} \Leftrightarrow$ e^-^2, $e^{-2\theta} \Leftrightarrow$ e^-^2\theta, $\cdots$ etc. What if we have a subscript (performed by the operator _) for example let us drop the symbolic text $x_{79}y_{12}$ on the last graph for which we execute the command **gtext('x_7_9y_1_2')** at the command prompt. After that go to the figure window, choose any position in the plot to drop the subscript text, and click the left button of mouse to find the text as in figure C.1(f). Multiple subscript writing follows the syntax similar to that of the superscript. The reader may need to know the MATLAB codes for frequently encountered Greek symbols, which are presented in Table C.1.

✦ How to include peripheral features on a drawn graph?

By peripheral feature what we mean is attaching some label once a graph has been drawn. Concerning the figure 4.1(b), the horizontal axis labeling is **x variation in cm**. How did we add that? There are two ways - exercising the command **xlabel** or clicking the menu by mouse. After the graph has been drawn, exercise the following:

>>xlabel('x variation in cm') ↵

You will find the horizontal string attached like figure 4.1(b). Any string must be inserted under quote. Similarly for the vertical axis we use the command **ylabel**. For example the figure 4.1(b) displayed text is attached by:

>>ylabel('f(x, 3 cm, -2 cm) in Volt') ↵

What about the 3D graphics label like figure 4.2(c)? We use analogous command **zlabel**. For the figure 4.2(c) displayed label we must have executed:

>>zlabel('f(x, y, -2 cm) in Volt') ↵

For the other option: find **Insert** at the menu bar of figure window, click the menu **Insert**, find **xlabel**, **ylabel**, **zlabel**, etc in a pulldown menu, and click the intended one. After that the cursor will be blinking at intended axis. Enter your text without a quote e.g. **x variation in cm** for the **xlabel**.

Appendix D

Creating a function file

A function file is a special type of M-file (section 1.1) which has some user-defined input and output arguments. Both arguments can be single or multiple. The first line in a function file always starts with the reserve word **function**. A function file must be in your working path or its path must be defined in MATLAB. Depending on problem, a function file is written by the user and can be called from MATLAB command prompt or from another M-file. For convenience, long and clumsy programs are split into smaller modules and these modules are written in a function file. The basic structure of a function file is as follows:

MATLAB Prompt   function file

$$>> g = \text{call } f \qquad \Longrightarrow \qquad \underbrace{g(y_1, y_2, ...y_m)}_{\text{output arguments}} = \underbrace{f(x_1, x_2, x_3, ...x_n)}_{\text{input arguments}}$$

We present following examples for illustration of function files keeping in mind that the arguments' order and type of the caller and function file are identical.

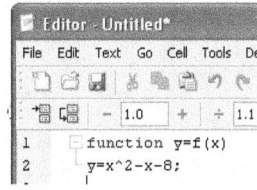

Figure D.1(a) Single input – single output function file

⊟ **Example 1**

Let us say the computation of $f(x) = x^2 - x - 8$ is to be implemented as a function file. When $x = -3$ and $x = 5$, we should be having 4 and 12 respectively.

The vector code (appendix A) of the function is **x^2-x-8** assuming scalar **x** and obviously the **x** is for x. We have one input (which is x) and one output (which is $f(x)$). Open a new M-file editor, type the codes of figure D.1(a) exactly as they appear in the M-file, and save the file by the name **f**. The assignee **y** and independent variable **x** can be any variable of your choice, which are the output and input arguments of the function respectively. Again the file and function name **f** can be any user-chosen name only the point is the chosen function or file name should not exist in MATLAB. Let us call the function **f(x)** to verify the programming as shown in the right side text box. You can write dozens of MATLAB executable statements in the file but whatever is assigned to the last **y** returns the function **f(x)** to **g**. Writing the = sign between the **y** and **f(x)** in the function file is compulsory.

Calling for example 1:

$>>$g=f(-3) ↵   ← call $f(x)$ for $x = -3$

g =

    4

for $x = 5$,

$>>$g=f(5) ↵

g =

    12

⊟ Example 2

Example 1 presents one input-one output function how if we handle multiple inputs and one output? The input argument variables are separated by commas in a function file. A three variable function $f(x_1, x_2, x_3) = x_1^2 - 2x_1 x_2 + x_3^2$ is to be computed by a function file. The input arguments (assuming all scalar) are x_1, x_2, and x_3 and

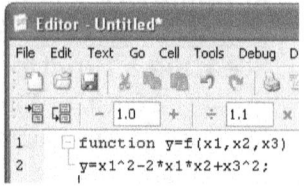

Figure D.1(b) Multiple inputs – single output function file

the output argument is the functional value of the function. The x_1 is written as **x1**, and so is the others. Follow the M-file procedure of example 1 but the code should be as shown in figure D.1(b). Let us inspect the function (with the specific $x_1 = 3$, $x_2 = 4$, and $x_3 = 5$, the output value of the three variable function must be $f(3,4,5) = 3^2 - 2 \times 3 \times 4 + 5^2 = 10$) as presented in the text box below.

```
Calling for example 2: when input arguments are all scalar:
>>g=f(3,4,5) ↵      ← calling f(x₁, x₂, x₃) for x₁=3, x₂=4, and x₃=5

g =
      10
Calling for the example 2: when input arguments are all column matrix:
>>x1=[2 3 4]'; ↵   ← x₁ values are assigned to x1 as a column matrix
>>x2=[-2 2 5]'; ↵  ← x₂ values are assigned to x2 as a column matrix
>>x3=[1 0 3]'; ↵   ← x₃ values are assigned to x3 as a column matrix
>>f(x1,x2,x3) ↵    ← calling f(x₁, x₂, x₃) using column matrix input arguments

ans =
      13
      -3
     -15
```

The **function** not only works for the scalar inputs but also does for matrices in general for example a set of input argument values are $x_1 = \begin{bmatrix} 2 \\ 3 \\ 4 \end{bmatrix}$, $x_2 = \begin{bmatrix} -2 \\ 2 \\ 5 \end{bmatrix}$, and $x_3 = \begin{bmatrix} 1 \\ 0 \\ 3 \end{bmatrix}$ for which the $f(x_1, x_2, x_3)$ values should be $\begin{bmatrix} 13 \\ -3 \\ -15 \end{bmatrix}$ respectively.

The computation needs the scalar code (appendix A) of $f(x_1, x_2, x_3)$ regarding x_1, x_2, and x_3. The modified second line statement of the figure D.1(b) now should be **y=x1.^2-2*x1.*x2+x3.^2;**. On making the modification and saving the file, let us carry out the commands which are placed in above text box of this page too. If it is necessary, the output can be assigned to user-supplied workspace variable v by writing v=f(x1,x2,x3) at the command prompt. The return from the function file also follows the same input matrix order. If the input arguments of f(x1,x2,x3) are rectangular matrix, so is the output. Input arguments of a function file do not have to be mathematics symbol. Suppose x_1 =ID, x_2 =Value, and x_3 =Data, one could have written the first and second lines of function file in figure D.1(b) as function y= f(ID,Value,Data) and y=ID.^2-2* ID.*Value+ Data.^2; respectively.

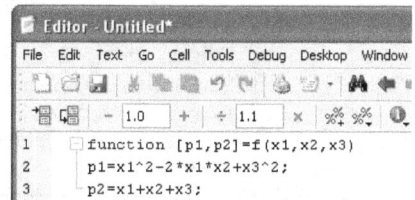

Figure D.1(c)   Function file for three input and two output arguments

⬚ Example 3

To illustrate a multi-input and multi-output function file, consider that p_1 and p_2 are to be found from three variables x_1, x_2, and x_3 (all are scalars) employing the expressions $p_1 = x_1^2 - 2x_1 x_2 + x_3^2$ and $p_2 = x_1 + x_2 + x_3$ whose function file (type the codes in a new M-file editor and save the file by the name f) is presented in figure D.1(c).

Choosing x_1 =4, x_2 =5, and x_3 =6, one should get p_1 =12 and p_2 =15 for which right side text box commands are conducted at the command prompt. More

Function file calling for the example 3:
>>[p1,p2]=f(4,5,6) ↵ ← calling the function file f for p_1 and p_2 using x_1 =4, x_2 =5, and x_3 =6
p1 =
12
p2 =
15

than one output arguments (which are here p_1 are p_2 and represented by p1 and p2 respectively) are separated by commas and placed inside the third bracket following the word function in figure D.1(c).

When calling a function, the output argument writing is similar to that of the function file (that is why we write [p1,p2] as output arguments at the command prompt). The output argument variable names do not have to be p1 and p2 and can be any name of user's choice. If there were three output arguments p_1, p_2, and p_3, the output arguments in the function file would be written as [p1,p2,p3] and their calling would happen in a like manner.

Note: We saved different files by the same name f just for simplicity and maintaining unifying approach. By this action any previously saved file f disappears. You may save the function file by other name like f1 and call accordingly e.g. the first line of figure D.1(c) would be function [p1,p2]= f1(x1,x2,x3) and calling would take place as [p1,p2]=f1(4,5,6) for the last illustration. For later version of MATLAB, add reserve word end at the end of every function file.

References

›› ›› **Electromagnetics Fundamentals** ›› ››

[1] Matthew N. O. Sadiku, *"Elements of Electromagnetics"*, Second Edition, 1995, Oxford University Press, New York.

[2] Constantine A. Balanis, *"Antenna Theory – Analysis and Design"*, 1982, John Wiley & Sons, Inc., New York.

[3] Jin Au Kong, *"Electromagnetic Wave Theory"*, 1986, John Wiley & Sons, New York.

[4] Nannapaneni Narayana Rao, *"Elements of Engineering Electromagnetics"*, Third Edition, 1991, Prentice Hall, Englewood Cliffs, New Jersey 07632.

[5] Steven E. Schwarz, *"Electromagnetics for Engineers"*, 1990, Saunders College Publishing, Philadelphia.

[6] Jack Vanderlinde, *"Classical Electromagnetic Theory"*, 1993, John Wiley & Sons, Inc., New York.

[7] Edward J. Rothwell and Michael J. Cloud, *"Electromagnetics"*, 2001, CRC Press LLC, Boca Raton.

[8] E. J. Post, *"Formal Structure of Electromagnetics – General Covariance and Electromagnetics"*, 1962, North Holland Publishing Company, Amsterdam.

[9] George Tyras, *"Radiation and Propagation of Electromagnetic Waves"*, 1969, Academic Press, Inc., New York.

[10] Branko D. Popović, *"Introductory Engineering Electromagnetics"*, 1971, Addison-Wesley Publishing Company, Reading, Massachusetts.

[11] Max Mason and Warren Weaver, *"The Electromagnetic Field"*, 1929, Dover Publications, Inc., New York.

[12] C. M. Lerner, *"Problems and Solutions in Electromagnetic Theory"*, 1985, John Wiley & Sons, New York.

[13] D. S. Jones, *"Methods in Electromagnetic Wave Propagation"*, Volume 2, 1987, Clarendon Press, Oxford.

[14] J. Caldwell and R. Bradley, *"Industrial Electromagnetics Modelling"*, 1983, Martinus Nijhoff Publishers, The Hague, The Netherlands.

[15] John R. Reitz, Frederick J. Milford, and Robert W. Christy, *"Foundations of Electromagnetic Theory"*, Third Edition, 1979, Addison-Wesley Publishing Company, Reading, Massachusetts.

[16] Charles Herach Papas, *"Theory of Electromagnetic Wave Propagation"*, 1965, McGraw-Hill Book Company, New York.

[17] Mansour Javid and Philip Marshall Brown, *"Field Analysis and Electromagnetics"*, 1963, McGraw-Hill Book Company, New York.

[18] Daniel M. Dobkin, *"The RF in RFID – Passive UHF RFID in Practice"*, 2008, Newnes, Elsevier, UK.

›› ›› **MATLAB and/or Finite Difference** ›› ››

[19] W. Yu, X. Yang, Y. Liu, R. Mittra, and A. Muto, *"Advanced FDTD Methods: Parallelization, Acceleration, and Engineering Applications"*, 2011, Artech House, Norwood, MA.

[20] J. J. Tuma, *"Handbook of Numerical Calculations in Engineering"*, 1989, McGraw-Hill, Inc., New York.

[21] P. Monk, *"Finite Element Methods for Maxwell's Equations"*, 2003, Oxford University Press, Oxford.

[22] M. Nuruzzaman, *"Tutorials on Mathematics to MATLAB"*, 2003, AuthorHouse, Bloomington, Indiana.

[23] Duffy, Dean G., *"Advanced Engineering Mathematics with MATLAB"*, Second Edition, 2003, Chapman & Hall, CRC, Boca Raton.

[24] Shampine, Lawrence F. and Reichelt, Mark W., *"The MATLAB ODE Suite"*, 1996, The Math-Works, Inc., Natick, MA.

[25] Peter V. O'Neil, *"Advanced Engineering Mathematics"*, Third Edition, 1991, Wadsworth Publishing Company, Belmont, California.

[26] Serge Lang, *"Calculus of Several Variables"*, Second Edition, 1979, Addison–Wesley Publishing Company.

[27] B. G. Pachpatte, *"Integral and Finite Difference Inequalities and Applications"*, First Edition, 2006, Elsevier, Amsterdam, The Netherlands.

[28] A. R. Mitchell and D. F. Griffiths, *"The Finite Difference Method in Partial Differential Equations"*, 1980, John Wiley & Sons Ltd, New York.

[29] M. Nuruzzaman, *"Technical Computation and Visualization in MATLAB for Engineers and Scientists"*, February, 2007, AuthorHouse, Bloomington, Indiana.

[30] M. Nuruzzaman, *"Finite Difference Fundamentals in MATLAB"*, July 16, 2013, CreateSpace, South Carolina.

Subject Index

www.ingramcontent.com/pod-product-compliance
Lightning Source LLC
Chambersburg PA
CBHW070243190526
45169CB00001B/285